Medicine & Health Care in Early Christianity

Medicine & Health Care *in* Early Christianity

GARY B. FERNGREN

The Johns Hopkins University Press
Baltimore

Printed in the United States of America on acid-free paper
2 4 6 8 9 7 5 3 1

The Johns Hopkins University Press
2715 North Charles Street
Baltimore, Maryland 21218-4363
www.press.jhu.edu

Library of Congress Cataloging-in-Publication Data

Ferngren, Gary B.
Medicine and health care in early Christianity / Gary B. Ferngren.
p. ; cm.
Includes bibliographical references and index.
ISBN-13: 978-0-8018-9142-7 (hardcover : alk. paper)
ISBN-10: 0-8018-9142-6 (hardcover : alk. paper)
1. Medicine—History—To 1500. 2. Medicine—Religious aspects—History—To 1500.
3. Church history—Primitive and early church, ca. 30–600. I. Title.
[DNLM: 1. Christianity—history. 2. Religion and Medicine. 3. Delivery of Health
Care—history. 4. History, Ancient. WZ 51 F364m 2009]
R145.F47 2009
610—dc22 2008024674

A catalog record for this book is available from the British Library.

The purpose of medicine is to relieve suffering;
of religion to explain suffering or to help us accept it.

CONTENTS

ACKNOWLEDGMENTS

This volume addresses in a connected way the early Christian reception of Greek medicine and the origin and development of Christian medical philanthropy in the first five centuries of the Christian era. I began it while I held a resident fellowship at the Oregon State University Center for the Humanities. I continued my writing during a two-term sabbatical leave from Oregon State University and completed it with the support of Publication Grant No. 5G13 LM008687-02 from the National Library of Medicine. A Library Research Travel Grant from my university's Valley Library allowed me to use the resources of several European libraries.

For most of my career I have collaborated with Darrel Amundsen. It was he who first introduced me to the social history of ancient medicine, and our collaboration of more than thirty years has been a fruitful and personally rewarding one. I had hoped that we would be able jointly to undertake this study, but circumstances have not made that collaboration possible. I dedicate this book to him as a token of our friendship and with gratitude for his influence and generous assistance over a lifetime. Portions of chapters 4 and 5 were coauthored by him when they appeared in an earlier form.

I am grateful to Darrel Amundsen and Vivian Nutton for the painstaking care with which they have read the manuscript and to the Reverend Martin Emmrich for reading chapters 3 and 4. Without agreeing with everything I have written (in fact, each disagrees heartily on one or another point of view), they have saved me from many errors of fact and interpretation and led me to more nuanced views on several matters. Vivian Nutton's name appears frequently in these pages, more often by way of disagreement than I should have liked. I have benefited enormously from his scholarship and his willingness to engage in friendly dialogue regarding our differing views on the relationship of Christianity and medicine, which I hope I have not exaggerated. My daughters Heather Morton and Anne-Marie Nakhla greatly improved the style and logic of the narrative. Kate Zahnle-Hostetler provided valuable assistance in the final stages of preparing the manuscript. Jacqueline Wehmueller, executive editor of the Johns Hopkins University Press, has been both encouraging and patient in granting me two extensions in the submission of the manuscript. My wife, Agnes Ferngren, was diagnosed with Stage IV cancer while the manuscript was in progress, and she passed away before it was finished. It is beyond my ability to express adequately my deep gratitude to her for her love, support, and encouragement during thirty-six years of marriage. Without it this book could not have been written.

Portions of several chapters originally appeared elsewhere. Several of my earlier published views have changed over time, and those changes are reflected in this book. I acknowledge permission to incorporate the following, in whole or in part, with many modifications and adaptations, into this volume:

"Early Christianity as a Religion of Healing," *BHM* 66 (1992): 1–15. Reprinted with the permission of the Johns Hopkins University Press.

"Early Christian Views of the Demonic Etiology of Disease," in *From Athens to Jerusalem: Medicine in Hellenized Jewish Lore and in Early Christian Literature,* edited by S. Kottek, M. Horstmanshoff, G. Baader, and G. Ferngren (Rotterdam: Erasmus, 2000), 183–201. Reprinted with the permission of Erasmus Publishing Company.

"The Early Christian Tradition" (with Darrel Amundsen), in *Caring and Curing: Health and Medicine in the Western Religious Traditions* (New York: Macmillan, 1986; reprint, Baltimore: Johns Hopkins University Press, 1997), 40–64. Reprinted in part with the permission of Advocate Lutheran General Hospital.

"Philanthropy in Medicine: Some Historical Perspectives" (with Darrel Amundsen), in *Beneficence and Health Care,* edited by E. E. Shelp (Dordrecht: Reidel, 1982), 1–31. Reprinted in part with the kind permission of Springer Science and Business Media.

"The Imago Dei and the Sanctity of Life: The Origins of an Idea," in *Euthanasia and the Newborn: Conflicts Regarding Saving Lives,* edited by R. C. McMillan, H. T. Engelhardt Jr., and S. F. Spicker (Dordrecht: Reidel, 1987), 23–45. Reprinted in part with the kind permission of Springer Science and Business Media.

"The Organisation of the Care of the Sick in Early Christianity," in *Actes/Proceedings of the XXX International Congress of the History of Medicine,* edited by H. Schadewaldt and K.-H. Leven (Leverkusen: Vicom KG, 1991), 192–8. Reprinted with the permission of the editors.

"Lay Orders of Medical Attendants in the Early Byzantine Empire," in *Acts/Proceedings of the XXXI International Congress of the History of Medicine*, edited by R. A. Bernabeo (Bologna: Monduzzi Editore, 1988), 793–9. Reprinted with the permission of Monduzzi Editore SPA.

"The Early Christian Reception of Greek Medicine," in *Beyond "Reception": Mutual Influences between Antique Religion, Judaism, and Early Christianity*, edited by David Brakke, Anders-Christian Jacobsen, and Jörg Ulrich (Frankfurt: Peter Lang, 2006), 155–73. Reprinted in part with the permission of the editors and the publisher.

"Krankheit," in *Reallexikon für Antike und Christentum*, Band XXI, edited by Georg Schöllgen et al. (Stuttgart: Anton Hiersemann, 2006), cols. 966–1006. Translated and reprinted in part with the permission of Anton Hiersmann KG Verlag and the Franz Joseph Dölger-Institut der Universität Bonn.

I have not striven for consistency in citing the titles of ancient sources. Sometimes I use the Latin or Greek titles and sometimes (especially for well-known works) the common English translation (e.g., Augustine, *City of God* rather than *De civitate Dei*). For bibliographic data on texts and translations of the works of the church fathers that I cite or quote, see Johannes Quasten, *Patrology*, 4 vols. (1950–86), and Everett Ferguson, ed., *Encyclopedia of Early Christianity* (1990).

Medicine & Health Care in Early Christianity

CHAPTER ONE

Methods and Approaches

The intersection of medicine and the Bible, particularly in the Bible's passing references to illness and healing, has long fascinated medical professionals and lay readers alike. Many of the subjects that fall under the rubric of biblical disease and medicine have been repeatedly discussed.[1] What was the nature of biblical leprosy and how did it differ from modern leprosy? Why was it associated with ritual cleanness? Does the Holiness Code establish rules for public health that reflect an early Hebrew understanding of hygiene? Did a native medical tradition exist in Israel? Was Paul's "thorn in the flesh" a chronic disease or physical disability and, if so, of what nature? How does one explain demonic possession and miraculous healing as they are described in the healing narratives of the Gospels? And how does the modern reader account for Jesus's healings? Were they displays of miraculous power or examples of faith healing or suggestion? Was Jesus, as some modern scholars argue, a Mediterranean folk healer who employed contemporary methods of healing similar to those practiced by other itinerant healers?

Although I shall address some of these familiar questions, I wish in this study to focus on two broader and more comprehensive matters. First, what kind of healing did early Christians employ? Was it miraculous healing or healing by natural means (i.e., medicine)? Specifically, what were the attitudes of early Christians to medicine and physicians? One might infer from reading the Gospels that religious healing was normative among Christians in the New Testament, especially given the fact that it is featured so prominently in the Gospel accounts of Jesus's ministry. I shall try to correct this misapprehension. Chapter 2 describes the Greek medical theory and practice that formed the backdrop to the early Christian understanding of illness and its treatment, as well as the process by which Christians appropriated Graeco-Roman medicine. In chapter 3 I argue that early Christians accepted a naturalistic view of disease causation and rejected the belief that ordinary disease was caused by demons and that healing was effected by exorcism. Chapter 4 attempts to demonstrate that miraculous and religious healing played a minor role in the early church.

The second subject that I shall address in this volume is the origin of Christian medical philanthropy. Chapter 5 explores the ideological and theological background of Christian concepts of philanthropy that led to the creation of the hospital. I describe the differences between Christian and Graeco-Roman concepts of philanthropy as well as the process by which Christian medical charity grew out of specific elements of Christian theology. Chapter 6 traces the historical development of Christian medical philanthropy within the urban church (*ecclesia*) during the first three centuries of the Christian era, which paved the way for the sudden emergence of hospitals in the latter half of the fourth century.

Anyone who attempts to understand concepts of illness and healing in the early Christian world is confronted at once with the paucity of sources.[2] While we possess a good deal of Christian literature from the first through the fifth centuries, we find little that speaks directly of Christian views of healing. We have no medical treatises written by early Christians and no systematic discussion in the New Testament or other early Christian literature of medicine or physicians. Hence we must rely in large part on circumstantial evidence or on passing references in the sources in attempting to reconstruct early Christian healing practices and attitudes to medicine. But there is a deeper problem. The New Testament does not yield unambiguous answers to the kinds of questions we ask about sickness and healing because its authors' intention was not to provide information about them but to place them within the context of their intended purpose. "Sickness and healing," writes H. Roux, "are never approached in the Bible from the medical or scientific point of view, but always from the religious point of view, that is to say, from the viewpoint of the particular relationship which they create or make apparent between the sick person and God. It is not the nature of the sickness, its development or treatment, which receives attention, but the fact itself envisaged as an event significant of man's destiny or condition within the general perspective of the history of salvation."[3] Posing questions of sources that are separated from the modern interpreter by a considerable spatio-temporal as well as cultural and imaginative distance is a familiar problem to the scholar who seeks to interpret ancient texts. As a result, one is tempted to fill gaps in the evidence with parallels from other, ostensibly similar cultures, ancient or modern; base conclusions on arguments from silence; or give undue weight to the particular emphasis of an ancient author that may distort our understanding of a historical problem.

A case in point is the conspicuous role assigned in the Gospels to miracles of healing in the ministry of Jesus. Their prominence in the Gospels may suggest that early Christians relied heavily on miraculous healing for their ills. Moreover, from the frequent mention of demonic possession in close proximity to cases of disease that Jesus healed one may infer that popular belief commonly attributed disease to demonic etiology, and healing to exorcism or to other supernatural means. This inference is strengthened by the fact that references to miraculous healing are found in the early centuries of the church. It is natural to conclude that recourse to religious healing was normative for the Christian community, and this conclusion has been widely held. But the matter is not so clear cut. A careful reading of the New Testament will show that even in its pages not all healing was regarded as supernatural and not all Christians were healed of their infirmities. References to miraculous healing in the Christian literature of the second and third centuries are few. Then, quite suddenly in the late fourth century, frequent accounts of purported miraculous cures appear that indicate an increasing and widespread belief in supernatural forms of healing. But the evidence does not permit us to impose a schematic pattern of development on the early Christian community that leads in a straight path from Jesus's healing in the early first century to the reports of miraculous healing in the late fourth and fifth, this because of the lack of references to the latter during the second and third centuries. Simply put, miraculous healing cannot account for *all* or even *most* healing in the early Christian church. Furthermore, during the first five centuries of our era, changing cultural patterns of classical society appear to have had an impact on Christian thought and practice.

Erroneous assumptions about early Christian views of science, nature, and medicine have long bedeviled the study of the relationship between Christianity and medicine. These assumptions date from the Enlightenment, but they were popularized in North America by two influential works that appeared in the late nineteenth century: John William Draper's *History of the Conflict between Religion and Science* (1874) and Andrew Dickson White's *A History of the Warfare of Science with Theology in Christendom* (1896). White, the first president of Cornell University, argued that the early church had hindered the progress of science both by denigrating the investigation of nature and by subordinating observation and reasoning to the authority of scripture and theology. The Draper-White or "conflict" thesis, as it is called, became enormously influential in America in both popular and academic circles.[4] During much of the twentieth century it dominated the historical interpretation of the relationship of religion and science. Reflecting a positivist outlook, it viewed science as continually progressing and overcoming the entrenched antagonism of religious opinions, which invariably retreated before its advance. This view has been partially responsible for the widespread belief that early Christians were opposed to medicine, an unstated assumption that continues to underlie discussions of the place of medicine in the early church. It must be noted that science as a discipline did not exist in the ancient world, and although I use the word occasionally, I do so equivocally. The Greeks spoke of *natural philosophy,* which was one of the three branches of philosophy (moral and metaphysical were the other two). Natural philosophy dealt with what we term the physical sciences, that is, the behavior of physical objects or processes that are observable in nature.

In the last three decades of the twentieth century, several scholars have provided a systematic reevaluation of the conflict thesis. Two major contributions to the literature came from David Lindberg and Ronald Numbers (*God and Nature,* 1986) and from John Brooke (*Science and Religion: Some Historical Perspectives,* 1991). In the introduction to *God and Nature,* Lindberg argues for a historical picture that recognizes in the relationship between science and religion "a complex and diverse interaction that defies reduction to simple 'conflict' or 'harmony.' "[5] Brooke suggests that a "complexity thesis" is a more accurate model than the familiar conflict thesis.[6] There exists, in fact, little evidence for a conflict between medicine and early Christianity. In the second century Christian apologists (theologians who defended their faith philosophically against pagan critics) began the process of harmonizing Christian theology with Graeco-Roman philosophy. These church fathers hellenized Christianity by taking over elements of classical culture and incorporating them into a Christian worldview.[7] They often borrowed from Greek natural science and medicine to illustrate or buttress their theological arguments. Their views of medicine were positive, and they showed no reluctance to consider medicine as a gift of God even if instantaneous healing in answer to prayer was sometimes claimed by Christians. The arguments of Brooke and of Lindberg and Numbers have gained increasing acceptance among professional historians of science.

In describing healing in early Christianity I use several terms that require definition: *miraculous healing, magical healing,* and *natural healing.* The first, *miraculous healing* (synonymous with *religious healing* or *ritual healing*), is an extraordinary event that results from the intervention of a divine power beyond the normal course of nature.[8] Belief in the possibility of divine intervention in nature was nearly universal in antiquity among pagans, Jews, and Christians. The form and manner of miraculous healing within Christianity from the period of the New Testament through the fifth century were considerably varied. Typically (though not exclusively), miraculous healing was performed in the New Testament without the use of any medical means. After the close of the New Testament period (roughly at the end of the first century of the Christian era), Christians began, very gradually at first, to employ a variety of means, including prayer, rites of healing like the imposition of hands or anointing with oil, a sacramental act like baptism, or the relics of a saint. Although claims of miraculous healing were made throughout the period under our consideration, the magnitude and extent of these claims as contemporary phenomena varied considerably from time to time.[9] The ancients also commonly held a belief in a second kind of healing, *magical healing*, which involved the employment of amulets, incantations, or occult objects like herbs and gems that manipulated hidden preternatural forces within nature but outside its normal course.[10] Though encountered in Christian communities, magical practices were roundly condemned by Christian leaders. The third kind of healing, which was more commonly sought than miraculous or magical healing, was *healing by natural means*, which was therapy that ranged from the physician's repertoire to folk remedies, home cures, traditional treatments, and herbal recipes. It is difficult, if not impossible in practice, to separate fully the three categories. They overlapped, and none was thought incompatible with the others. All were used for healing in the ancient world, and the lines between them were sometimes as blurred in the ancient sources as in the minds of modern scholars—hence the need to avoid imposing modern categories on the data that one finds in the primary texts or drawing neat distinctions between them.[11] I employ two other terms, *supernatural* and *natural*, that are sometimes said to be inaccurate and anachronistic.[12] While a case can be made for substituting more precise terms, such as *divine/human* or *divine/nondivine,* or simply *marvelous,* which have the advantage of avoiding theological connotations, the terms *supernatural* and *natural* are so widely used that I believe they can be retained, when properly qualified, without serious misunderstanding.

It has become common for scholars to claim that no form of healing enjoyed primacy in what has been termed the "medical marketplace" of the ancient world. I believe that this claim somewhat overstates the case. Several kinds of medical practitioners were available: the empiric, for example, who largely treated symptoms, and the midwife, who delivered newborns. But beginning in the fifth century B.C. a new kind of medicine arose in Greece that was based on the application of theory to disease as a means of providing explanatory models. It is sometimes called *theoretical/speculative medicine*, sometimes *rational medicine*, sometimes *Greek* or *Hippocratic medicine.* Scholars have increasingly objected to calling it *rational medicine* both because it retained or incorporated elements of magic, religion, superstition, and folklore and because of the association of the concept of *rational* with Enlightenment-based modes of thinking. Yet, while it was not the only kind of medicine available in the classical world, over time it established its primacy among most other kinds. In the words of Philip van der Eijk, "Greek medicine, with its emphasis on explanation, its search for causes, its desire for logical systematisation, its endeavour to provide an epistemic foundation for prognosis and treatment, and especially its argumentative nature and urge to give account (*logos, ratio*) of its ideas and practice in debate, does show a distinctive character."[13]

Physicians held a variety of theories regarding the nature of disease and the healing properties of medicine, but they employed naturalistic therapies. It was a theoretical approach that differentiated them from other practitioners, such as those who administered home or folk remedies (and who might also employ naturalistic therapies) or religious healing.[14] Medical technique, moreover, dispelled the mystery of preternatural etiologies.[15] This medicine was Greek in origin, but it spread during the Hellenistic period (323–330 B.C.) throughout the Mediterranean world. By the first century of our era it was available in nearly every town and city throughout the Roman Empire. I refer to it as *Greek* or (in the Roman imperial period) *Graeco-Roman* or merely *secular medicine.* Its practitioners were known as *iatroi* (Gk.; *iatrinai,* female) and *medici* (L.; *medicae,* female). Although it has become fashionable to translate these terms as "healers" rather than as "physicians," I believe that the broader term ("healers"), taken over from medical anthropology, is somewhat misleading. Patients who sought healing from a *medicus* or an *iatros* could expect to receive both prognosis and treatment that relied on established medical theory and employed natural processes rather than magical or religious means. Nevertheless, standards of competence varied considerably. "People calling themselves *medici,*" writes Gillian Clark, "were an odd mixture of the very highly selected and the self-appointed, with a wide range of ability, qualifications, and prestige."[16]

Assumptions and Terminology

Certain assumptions underlie this study. The first is that we have in the New Testament a consistent and credible picture of Jesus's public ministry and the origins of the earliest Christian community. Since the origin of biblical criticism in the early nineteenth century, critical scholars have attempted to understand the Gospels in the light of Enlightenment assumptions. Taking for granted that the events described in the New Testament could not have occurred in the way they are described ("If miraculous, then unhistorical"), they have sought alternative explanations. In spite of nearly two centuries of the most painstaking effort to sort out the genuine words of Jesus from those allegedly created by the early church, there remains little substantive agreement among critical scholars. Every new generation finds it necessary to reopen the search for the historical Jesus in the light of changing assumptions —hence the novel, if often tendentious, reconstructions of his life that appear in every publisher's new list.[17] As a historian trained in reading classical texts, I find in the Gospels, as I do in the work of the classical Greek and Roman historians, promising material for the reconstruction of the events they describe.[18]

I find unconvincing, moreover, the view that the words of Jesus and the events of his brief career were radically modified by his followers after his lifetime, resulting in a discontinuity between his teachings and those of early gentile Christianity.[19] The assumption is widespread that the early church played a significant creative role in reshaping the earliest traditions regarding Jesus, with the result that his teachings came very quickly—within a generation—to be distorted, a process by which the "Jesus of history" was transformed into the "Christ of faith."[20] Those who hold this view have little confidence in the ability of the early Christian community to transmit accurately by oral or written tradition authentic memories of Jesus. They see in the Gospels little more than a mass of fragmentary and contradictory traditions. Hence what the New Testament preserves is the faith of the primitive church that has been imposed on the historical Jesus, from which we can recapture, by close textual analysis but with considerable difficulty, only fragments of his life and teaching. But why should we doubt the ability of the early church (a small and closely knit community) to preserve over one generation an accurate recollection of the events of Jesus's life and teachings? The personality of Jesus clearly made a strong impression on his followers, and it is a personality that is everywhere apparent in the Gospels, which are so easily distinguishable from the legendary accounts that grew up later. In fact, it was not until the second century that the mythmaking began, and we see its manifestation in the apocryphal and pseudepigraphical works of that period. Here, as elsewhere in dealing with historical sources, the brevity of time works in the opposite direction: the credibility of the Gospel writers is strengthened by the fact that they were under the scrutiny of eyewitnesses.

A second assumption that underlies this study is that the historical-philological method is the most productive way of understanding the evidence. Modern historians who have adopted the theories and models of the social sciences view theological or ideological factors in history as mere epiphenomena, preferring to employ cultural or material explanations of a sociological nature.[21] A social-anthropological approach to early Christian healing has been adopted by scholars such as John Pilch and Hector Avalos.[22] There can be, of course, no history without theory. Every historian seeks to adopt a framework that permits us to get inside the mind of ancient writers or to understand societies that are, by our standards, foreign. Historians have long drawn on the categories of social anthropology in their attempts to understand ancient cultures and societies. Used creatively, social anthropology provides perspectives that owe their insights to the fact that they borrow from the study of comparative cultures. Several classical historians (e.g., Geoffrey Lloyd, Peter Brown) have made use of them in supplementing, though not supplanting, more traditional historical approaches to ancient medicine and late antique society.[23]

But the insights of the social sciences provide pitfalls as well as benefits.[24] In attempting to impose an interpretive grid on early Christian understandings of health and disease, they sometimes minimize temporal and geographical distinctions. And they tend to relativize the ideas of the period under study while absolutizing those of their own. Ideologically laden with Western cultural assumptions, as they are, the methods of social anthropology need to be kept on a leash and used to complement—not to replace—the textual-philological historical method. Largely sociological explanations distort the reconstruction of historical events when they privilege social forces to the exclusion of the theological and philosophical concerns that play so prominent a part in our texts. The latter reflect the larger culture they inhabit and help us to understand some, but not all, of the factors that motivated the ancients to act as they did.[25] In areas such as the history of medicine and religion, where ideas and texts are so important, I find it difficult to neglect intellectual history. Hence I assume a mutual interaction in which ideas contribute to the shaping of cultural and social phenomena just as cultural and social phenomena contribute to the shaping of ideas.[26]

The past several decades have seen a sharp reaction to the largely positive view of the early church and its influence on society that dominated historical scholarship in the first half of the twentieth century.[27] The reaction has taken the form of a highly critical perspective, based largely but not wholly on poststructural and postmodernist theories, which reflect the influence of French scholars such as Roland Barthes, Jacques Derrida, and Michel Foucault. Poststructuralists argue that discourse reflects no reality other than its own and cannot be taken as an accurate portrayal of the events ("pretextual reality") it describes. Truth is not discovered but made, historical events not reconstructed but constructed. The meaning of any ancient text is not to be found in the author's world of thought but rather in the interpreter's own world, where it can be constructed in a manner that is free from the controlling function of the cultural or literary world that gave rise to it. Thus stories of miraculous healing become merely literary traditions of socially constructed reality and a means by which a group defines its own identity. This assumption allows plausible alternative, but often highly conjectural, counternarratives to be devised that are permitted by modern structures of reality. An example is literary-critical analysis in which one assumes that narratives of sickness and suffering have little to do with the external reality they describe but rather depict imagined states of "the suffering self" that can be imposed on society.[28] Discourse analysis is currently popular. It examines the rhetoric of early Christian literature in terms of its exploitation by a hierarchical institution that sought to justify its accumulation of power over society by creating an ideology of charitable concern. Sexually repressive and hypocritical in its alleged concern for the poor, according to this view, it employed philanthropy and charitable institutions like hospitals to gain access to public funds while exploiting the language of pity for the poor to victimize those in need. Michel Foucault has been highly influential in the formulation of this perspective.[29]

While I am indebted to the work of scholars who write from this point of view, I do not find it a convincing theoretical framework for understanding the changes that manifested themselves in late antiquity. Their desire to deconstruct the rhetoric of early Christian fathers, with its totalizing discourse, and their antipathy to institutional Christianity as a regime of power lead them to minimize the historical context of the rise of Christian charitable institutions. They do so by underplaying the widespread suffering that existed in the late Roman world and deconstructing the large body of rhetoric by which Christian leaders urged their followers to relieve that suffering and not merely to talk about it. Excessively attuned to the theoretical issues of poststructuralist criticism, some practitioners of this perspective adopt ahistorical modes and downplay human agency in a manner that sometimes approaches reductionism.[30] The notion of history as a process of cause and effect receives short shrift in a historical quest that focuses on modes of discourse as the nearest one can come to "real historical processes" or "extralinguistic realities."[31] Daniel Reff has argued, correctly, I believe, that it is against the backdrop of widespread epidemics, which began in the second century, that Christian discourse of suffering and philanthropy developed. He maintains that one cannot understand that discourse without appreciating how its growth reflected the epidemiological state of the culture in which it developed.[32]

In this study, I look at the evidence, insofar as it is possible for a modern scholar to do so, from the point of view of those to whom the texts were addressed.[33] Of course, it is impossible to do this adequately. When ancient texts describe outdated theories of disease, or attribute disease to a demonic etiology, or assert that plagues and disease were sent by gods who had been offended and needed to be propitiated, one realizes that the modern mind will never fully grasp the way in which the ancients created their particular view of reality. This study makes a special attempt to understand the theological constructs that lay behind pagan, Jewish, and Christian medical practices. To ignore theological questions by substituting comparative cultural explanations or socially constructed discourse, in my opinion, often obscures rather than clarifies the issues. All historical reconstructions are tentative and subject to continual modification not merely in the light of new evidence but also in the light of changing assumptions and new questions that are posed to the evidence. My goal is to reach a historical understanding of a particular religious community within an ancient Mediterranean culture. That task involves appreciating Jewish and Christian theological views that underlay ancient concepts of health and disease. It means wrestling with ancient texts. And it persistently strives to avoid the assumption that modern Western cultural values or current theoretical constructs constitute the bar before which all other values are judged. The modern age is a historical period like any other, limited in its perspectives by time and culture and subject to the constraints of its own zeitgeist. Understanding that we, too, have historical and cultural limitations forces us to view the past in a manner that is neither patronizing nor disparaging but appreciative of the power of ideas and practices that we do not always share or fully understand.

A third assumption that underlies this study is that early Christians were citizens of the world in which they lived and that they held many of its cultural presuppositions. The earliest Christians were Palestinians whose beliefs were rooted in a first-century Jewish culture.[34] Non-Jewish Christian converts did not necessarily abandon many of the views that they had held earlier as pagans. As subjects of the Roman Empire they operated within a milieu that was both synthetic and cosmopolitan. Their religious values undergirded their worldview. But their understanding of medicine reflected the values that had permeated the Mediterranean world, Jewish, Graeco-Roman, and Christian. Although there were Jewish physicians, no distinctly Jewish medicine existed. Jews had adopted Greek medicine, while doubtless retaining some elements of their traditional healing culture. Nor did a specifically Christian medicine exist at any time. There was only Greek (or, after the first century B.C., Graeco-Roman) medicine, and adherents of all religions accepted it to some degree.[35] Of course, Christian religious values raised moral, ethical, and theological questions regarding medicine and healing. But a first-century physician like Luke, who traveled throughout the Roman Empire with the apostle Paul, would have felt no cultural boundaries; the medicine he practiced reflected the training and medical views that he shared with many colleagues elsewhere.[36] Hence an understanding of the larger classical world casts much light on early Christians, who cannot be studied apart from the Graeco-Roman culture they inhabited. If we wish to understand medicine and healing in the early centuries of our era, we must explore the commonly accepted concepts of disease, the practice of medicine, and the social position and attitudes of physicians, as well as the respective theologies and cultural boundaries that informed their acceptance of Greek medicine. Knowledge of this larger world of Graeco-Roman culture provides a contextual understanding of Christian approaches to medicine and healing.[37]

Two matters remain to be addressed. The first is the definition of Christianity that I employ. In the past generation there has been a reevaluation of what constituted early Christianity, given the recovery of previously lost writings that have afforded us a broader knowledge of movements, usually termed sectarian or heretical, that grew up on the fringes of mainstream Christianity.[38] The most obvious example is Gnosticism, which was, until the publication of the Nag Hammadi library discovered in 1945, a little-known religious movement that in the second century became a rival to mainstream Christianity. Gnosticism comprised several esoteric religious sects, some that claimed to be Christian and some that did not, which taught that salvation comes through knowledge (*gnosis*). I have largely restricted my discussion to the incarnational Christian movement that is represented in the New Testament and the writings that have come down to us through the fathers of the church, which came to be defined in the early Christian creeds as orthodox.[39] The one exception is Montanism, a second-century Christian movement that arose in Asia Minor, which I examine in chapter 4 in the context of purported claims of miraculous healing. While a study of healing and health care as practiced by what came to be termed heterodox forms of Christianity is desirable, they deserve separate treatment. I use several commonly accepted terms to describe what W. H. C. Frend calls "the Great Church": *catholic, orthodox, mainstream.*[40]

My second point concerns terminology. I use the conventional terms *Christian* and *pagan* while acknowledging that they encompass a broad spectrum of belief and unbelief. The term *pagan* is culturally offensive to some today. Yet there is no other word that has the same historical associations in describing those who worshiped the old Greek and Roman deities. Originally the word meant simply one who lived in the country, a rustic. It began to be applied in the early fourth century to people (many of whom lived in the countryside) who continued to observe the traditional Roman religious ceremonies instead of accepting the new religion of Christianity, which grew up in the cities. According to Robin Lane Fox, "By modern historians, pagan religion has been defined as essentially a matter of cult acts. The definition has an obvious aptness. Pagans performed rites but professed no creed or doctrine. They did pay detailed act, especially by offering animal victims to their gods, but they were not committed to revealed beliefs in the strong Christian sense of the term."[41] It hardly needs to be said that in employing the word *pagan* I intend no pejorative connotations. Some writers have protested against the use of a binary opposition of *Christian* and *pagan* on the ground that the distinctions were not always as sharply defined as we make them today.[42] A good deal of discussion has been devoted to the definition and nature of conversion,[43] pagan survivals, and assimilation in late antiquity.[44] The evidence suggests that, in their relations and debates, pagans and Christians were not always in conflict, that they often shared a common culture, and that the exclusive language of warfare between them is misleading. The words of David Lindberg, used in a different context (the historical relationship of Christianity to science), are applicable here: "What we find is an interaction exhibiting all of the variety and complexity with which we are familiar in other realms of human endeavor: conflict, compromise, understanding, misunderstanding, accommodation, dialogue, alienation, the making of common cause, and the going of separate ways."[45] Yet the terms describe significant differences that existed not only in religion but, in important respects, in culture and worldview. For that reason, and not merely because they are traditional, I have retained them.

CHAPTER TWO

The Christian Reception of Greek Medicine

It is the thesis of this book that Christians of the first five centuries held views regarding the use of medicine and the healing of disease that did not differ appreciably from those that were widely taken for granted in the Graeco-Roman world in which they lived. They did not attribute most diseases to demons, they did not ordinarily seek miraculous or religious cures, and they employed natural means of healing, whether these means involved physicians or home or traditional remedies. In this chapter I shall describe the understanding of disease and the kinds of medical treatment that were current in the classical world during the earliest centuries of Christianity. We shall begin by considering ancient concepts of disease and therapeutic approaches of Greek physicians who belonged to one or another medical sect. Since Christianity arose within the context of Palestinian Judaism, we shall trace the history of the pre-Christian Jewish understanding of healing and disease. We shall explore the attitudes toward Greek medicine of the second-century Christian apologists, focusing on those like Tertullian, Origen, and Tatian, whom some scholars have thought to be hostile to medicine. I shall attempt to demonstrate that these apologists' understanding of medicine was compatible with recourse to the generally accepted medical knowledge of the Roman imperial period. Finally, we shall examine in some detail the views of the fourth-century rhetorician Arnobius, who, as a recent convert to Christianity, offers an example of an intellectual Christian who was anxious to denounce pagan views that he had only recently abandoned but who considered medicine a gift of God and hence appropriate for Christians to use.

Concepts of Disease

Every language employs a variety of words, some precise and some imprecise, to express the manifestations of physical disability and dysfunction. When imprecise terms are used to indicate that someone is ill, the language is usually not scientific and does not denote a particular medical condition. One often employs general terms to express merely the perception that one is not (in the commonly accepted understanding of the term) in good health. Hence words like *illness* and *sickness* carry with them a social or cultural rather than a scientific medical connotation. Among medical anthropologists *disease* is usually taken to describe a pathological condition, while *illness* denotes a subjective feeling of physical discomfort.[1] But even when a specific (i.e., medical) vocabulary is employed, if it is symptomatic rather than pathological, it normally reflects the particular culture that created it rather than indicates a scientific description. And when the vocabulary of disease becomes pathologically specific, it reflects a particular nosology or medical model that is culture-specific and cannot be taken as necessarily consistent with modern nosological constructs.[2]

Comparative Nosologies

Medical anthropologists have developed several conceptual frameworks for classifying the etiologies of disease.[3] One widely used scheme, developed by George Foster, divides disease etiologies into *personalistic* and *naturalistic* causes. The former views disease as the result of intervention by human or supernatural agents, while the latter views it as caused by natural forces. A second scheme, that of Claudine Herzlich, posits disease as either *exogenous,* which views its cause as external, or *endogenous,* which views its cause as internal. The causes may be either natural or supernatural. A third system is that of Margaret Lock, who classifies etiologies as either *ontological* or *physiological.* The former views disease as caused by an outside agent that attacks the body, while the latter considers disease an imbalance within the body. The three systems are not mutually exclusive, and they sometimes overlap. Indeed, they often coexist within the same culture. The ontological theory views disease as both specific and contagious or communicable.[4] According to this model, disease is an invasive entity that is destroyed or counteracted by medicine. Since the nineteenth century, ontological nosologies have been dominant in the Western world.[5] In the classical world physiological nosologies were dominant.

The relation of a group of symptoms to a specific disease or disease agent is a recent medical concept. Greek medical authors regarded diseases not as separate entities but as groupings of symptoms that affected individuals and could be described and classified.[6] Symptoms were indicators of changes that took place in the constitution of the individual patient. A disease represented a deviation from the normal state of a patient. Although the "antecedent" or external cause might vary, a disease led (though not among the Methodists) to a humoral disorder ("standing condition"), which produced a malfunction of the body. Thus a cold wind might create an excess of phlegm in the stomach, which in turn would produce indigestion. The susceptibility of an individual to disease determined how he or she responded. Some people were more disposed to indigestion than others, even though both groups shared a common climate, weather, or diet. But it was the standing condition, such as an excess of phlegm in certain individuals, not the wind or the antecedent predisposition to chills, that led to indigestion. Treatment consisted of creating or restoring the proper humoral balance in each individual—for example, by reducing the excess of phlegm in the stomach. But the proper balance differed from individual to individual. Hence ancient physicians were illness-oriented, and they viewed and treated the patient as a whole person rather than the disease itself.[7]

Some diagnoses found in ancient medical literature (e.g., of tetanus, epilepsy, mumps orchitis) appear relatively clear in terms of modern nosologies. But as we learn more about rare and unusual causes of conditions, we become more cautious about making diagnoses based only on the description of a few symptoms.[8] Diseases, moreover, change over time in frequency, severity, clinical aspect, and epidemiological peculiarities through variation and adaptation under changing environments by a process of natural selection.[9] This is true especially of infectious diseases and those conditions that involve changes caused by humans.[10] It is likely that many disease entities have been transformed over the centuries and that whatever disease (or combination of diseases) constituted an ancient epidemic like the plague of Athens, it cannot be precisely identified with any disease that exists today, in spite of the detailed description of the symptoms given by Thucydides.[11] Hence retrospective diagnosis must always remain problematic.[12] But epidemics were sporadic, whereas pneumonia, pleurisy, and pulmonary diseases of all kinds are frequently described by medical writers. Other factors, however, complicate attempts to identify ancient diseases in terms of modern nosologies. One is the general lack of precision employed by ancient writers in describing disease symptoms. The term *phthisis* covered a spectrum of consumptive diseases that included pulmonary tuberculosis.[13] The generic term *fever* (Gk., *puretos*; L., *febris*), which in modern medicine is merely a symptom, was used by Greek medical writers to refer to any disease that produced a high temperature. Fevers were often described in terms of their recurrent patterns of remission and return. Remittent or intermittent malaria, which was endemic in several places in Greece (e.g., Boeotia, Macedonia) and Italy (e.g., Campania), provided the model, since it was in malaria that a recurring pattern was most commonly observed.[14]

Ancient diseases were frequently named after the part of the body that was affected. But no standard terminology existed among medical authors, and the fact that different authors used widely differing terms for the same condition sometimes resulted in confusion. Classical writers, moreover, were selective in their description of symptoms, often omitting those that are crucial to differential diagnosis. This can be seen in the extensive collection of individual cases described in the Hippocratic *Epidemics*, whose signs and symptoms (like those described by Thucydides) were intended to enable the reader to identify a particular condition in the future.[15] The fact that the same symptoms are common to different diseases renders the specific identification of a disease problematic. The symptoms and impact of a disease can vary, moreover, with nutritional conditions and diet. A further complicating factor (especially in epidemics like the plague of Athens and the Black Death) is the possibility that we are dealing with two disease entities that together have been the cause of especially virulent epidemics.[16] Rhetorical and stylistic features of our sources can also hinder our understanding of the specific nature of a disease, a phenomenon that is especially marked in describing the death of famous figures.[17] Classical writers were sometimes influenced by classical descriptions of earlier epidemics and borrowed both their style and substance, including descriptions of the symptoms. Thus Procopius's description of the plague of Justinian relies in more than a merely formal sense on Thucydides' description of the plague of Athens centuries earlier.[18]

Perhaps the best-known examples of ancient disease symptoms whose identification is problematic are leprosy and syphilis. Biblical "leprosy" as described in both Old (Lev. 13:47–59, 14:33–53) [19] and New Testament (Mk. 1:40–42 [= Mt. 8:2–3, Lk. 5:12–14]; Lk. 17:12) narratives appears to encompass a spectrum of lesions that is broader than Hansen's disease. The latter is caused by a specific microorganism, *Mycobacterium leprae*, and its symptoms are a thickening of the skin with patches of discoloration and loss of sensation owing to the deterioration of the peripheral nervous system.[20] But the Hebrew word *za'arath* (Gk., *lepra*) describes a scaly condition that affected not merely the skin but also clothing and the walls of a house and probably included psoriasis, ringworm, and various fungal conditions. The vagueness of the symptoms in our sources makes it difficult to differentiate the various diseases that were subsumed within the term, but elephantiasis was among them. The disease was attributed by medical writers to an excess of black bile, and viper meat, either taken internally or applied externally, was frequently prescribed to treat it.[21] Physical disfigurement may or may not fully explain the revulsion that the condition produced, and the reason for the association of leprosy with ritual defilement in Hebrew culture remains unclear. While social ostracism and social terror accompanied the disease in Graeco-Roman society, it did not result in the ritual defilement and pollution that it produced in Hebrew society.[22] The relationship of syphilis to its ancient precursors is similarly problematic.[23] Venereal syphilis is one of several diseases that are caused by bacteria of the genus *Treponema* (the others are pinta, yaws, and endemic syphilis). It has long been debated whether it antedated Columbus's discovery of the Western Hemisphere or was brought to Europe by returning European explorers. The osteological evidence before the sixteenth century is sparse. While lesions of the later stages of the modern disease have been found in a few ancient skeletons, it is uncertain whether they are of a venereal or a nonvenereal infection. If the disease was present in antiquity, it was probably in a different form than that found in modern times.

Environmental Factors in Disease

The author of the Hippocratic *Airs, Waters, and Places* attributes disease to climatic conditions like seasonal variation, winds, and temperature, as well as to geographical and demographic factors. The Greeks associated disease with certain seasons (pleurisy in winter, jaundice in summer), while thinking that different localities were susceptible of different diseases. Diarrhea (common in summer) and colds (frequent in winter) were similarly accounted for by seasonal effect on the humoral cycle.[24] The Greeks had observed from early times that certain areas, like low-lying marshes or slow-moving rivers, were unhealthy and that certain diseases, like malaria, were endemic in those regions. Epidemic diseases were frequently attributed to hot stifling air or to vapors ("bad air") that made the air poisonous. The Hippocratic treatise *On the Nature of Man* attributes the cause of bad air to contamination by exhalation or excretion of some kind. In contrast, cool gentle breezes and well-ventilated houses were considered to be healthy, perhaps on the basis of observing conditions that were free of malaria. Medical writers prescribed changes in diet and regimen (including breathing patterns) to strengthen the individual's ability to overcome harmful air. The classical world believed that *some* diseases could be passed from person to person by touch or proximity, but ancient medical writers never developed a theory of contagion.[25] "What passes is an emanation, an effluxion, a breath, a poison, a putrid effusion, an excrement, or a miasma," but it was the susceptibility of individuals that resulted in disease infection.[26] Epidemic diseases were not generally thought to be contagious.[27]

Greek Medicine
Empirical and Theoretical Medicine

As early as the time of Homer there existed in Greek society physicians (*dēmiourgoi*) who treated wounds and set bones. They were itinerant craftsmen who relied on skill and observation and learned their craft through apprenticeship.[28] They could identify symptoms and apply traditional remedies. But they did not understand disease in general terms; nor were they able to frame theories that could be applied to particular cases. In the fifth century B.C. some physicians began to explain disease in terms of natural causation. They borrowed theories of the nature of health and disease from the physiological speculations of pre-Socratic philosophers, who had substituted naturalistic for mythological explanations of the natural world. Thus Alcmaeon of Croton (c. 500 B.C.) maintained that health represented a balance of such opposites as the dry, the wet, the hot, the cold, the sweet, and the bitter. Illness resulted from an imbalance of these contrary forces. Dietary therapy prescribed foods whose characteristics helped restore the body's balance (e.g., hot to balance cold, dry to balance wet). Perhaps the best-known theory was based on the supposed existence of humors, which physicians borrowed from Empedocles (fl. 444–441 B.C.), according to which the body contained four fluids (blood, phlegm, yellow bile, and black bile), by analogy with matter, which was composed of four "elements" (earth, air, fire, and water). Most doctors believed that health resulted from a harmonious balance of the humors of the body (a pathological theory that was inspired by political observation, as, e.g., in Plato's *Timaeus*),[29] while disease was caused by a disturbance or imbalance of humors. The treatment of disease was directed to restoring the harmony of the humors through "coction," a mixing of the humors that usually involved the combining of dietetics with cathartic therapies (e.g., emetics, purging, or bloodletting). This approach, which consisted of applying treatments that were similar to the condition that they sought to cure, undergirded much ancient medicine. Since the humors tended to an equilibrium of their own accord, Hippocratic physicians preferred to leave those people whose condition they could not alleviate to the *vis medicatrix naturae* ("healing power of nature") and the naturally recuperative powers of the body. Even within the Hippocratic Corpus there were variants (e.g., *On Breaths* attributes disease to "breaths"), but the theory of the four humors, set forth in *On the Nature of Man,* eventually emerged as the dominant disease model largely because of its adoption by Galen (A.D. 129–c. 199/216) in the second century. This physiological (as opposed to merely empirical) approach to medicine rested on a theoretical underpinning that largely omitted religious or magical factors in explaining the etiology of disease while not categorically rejecting religion or divine intervention.[30] It is seen in the more than sixty Hippocratic treatises that date from the late sixth century to the third century B.C., most of them from the late fifth or fourth century B.C., with a few dating much later. Thus the treatise *The Sacred Disease* rejects a divine explanation for epilepsy, which was termed the "sacred disease" because it was widely attributed to particular gods or to divine forces. The anonymous author argues that, like other diseases, it has a natural cause, his own explanation being highly speculative.[31] "There is no need to put the disease in a special class and to consider it more divine than the others; they are all divine and all human. Each has a nature and power of its own; none is hopeless or incapable of treatment" (ch. 21).

Greek Medical Sects

Greek theoretical medicine, with its assumption of natural causation of disease, spread during the Hellenistic period throughout the Mediterranean world and eventually supplanted in many locales the more primitive medicine of the empirics or folk healers that had preceded it, though the latter remained a fixture in every community. Most physicians received their training by apprenticeship. Loosely organized communities of physicians, scattered throughout the eastern Mediterranean at sites like Kos and Knidos and in the West in Sicily and Magna Graecia (southern Italy), attracted those who wished to study medicine.[32] From the third century B.C., Rome became a part of the Hellenistic world, and though it conquered the Greeks in a series of wars that concluded in 146 B.C., Greek culture spread rapidly to Rome, and it included medicine. The first person said to have practiced medicine in Rome as a distinct profession was the Greek Archagathus, who settled in Rome in 219 B.C.[33] He was treated with great respect by the Romans, but his excessive use of surgery and cautery made him unpopular, and he came to be called *carnifex* ("the executioner"). In spite of opposition from conservatives who were hostile to Greek medicine, such as Cato the Elder (234–149 B.C.), Greek physicians came quickly to enjoy acceptance at Rome.[34] But traditional Roman medical practices, consisting of magic and folk medicine, remained popular even after the large immigration of Greek physicians into Italy in the second and first centuries B.C.; perhaps not surprisingly, they appealed to old-fashioned Romans for centuries. This is illustrated by the compilation of folk and magical remedies made by Pliny the Elder (A.D. 23/24–79) in his *Natural History*; indeed, they long continued to be used. Romans also continued to consult soothsayers when ill.[35]

Widespread disagreement existed regarding the theoretical basis of disease. Many physicians accepted the theories of one of the medical sects (*haireseis*) that arose in

Hellenistic times, the chief of which were the Dogmatists and the Empiricists.[36] The Dogmatic sect held that a knowledge of the internal organs ("hidden causes") was necessary before treatment could be administered; hence Dogmatic physicians practiced dissection and even, on rare occasions, vivisection.[37] The Empiricists, in contrast, refused to seek hidden causes, which they regarded as conjectural, observing physical symptoms instead. They avoided theory and based their practice on experience, which revealed those treatments that had succeeded in the past and those that had failed.[38] A third major sect was Methodism, which arose during the Roman period. Methodists rejected humoral pathology and a theoretical approach to medicine altogether. They argued that all diseases shared common conditions (the constricted, the lax, and the mixed) that were easily recognizable, since the whole body rather than merely the affected part exhibited the signs of disease. Once the conditions were identified, it was easy to determine the treatment.[39] All three of these schools were especially prominent in the first century B.C. and the first century of our era. But the medical sects were never monolithic, and they displayed considerable variation in both theory and practice. There were, moreover, many physicians who were not connected with any sect. But no unanimity existed either as regards the nature of disease or how it was to be cured.

Therapeutics

Physicians in the classical world emphasized individual treatment. The treatment of disease usually took the form of a holistic approach that involved the use of regimen, pharmacology, and surgery. Preventive medicine was regarded as the best way to maintain health. This was accomplished by means of a detailed classification of food and drink based on the properties of each, whether strong, weak, dry, moist, cool, hot, constricting, or laxative. Sleep, sexual activity, and exertion were also regulated. Treatment was adjusted to the individual, and a physician was expected to know his patients well so that he would be able to adjust their treatment in order to maintain a balance of the constituent elements of the human body. Drugs (*pharmaka*) were commonly administered but with uncertain results.[40] Cautery and surgery were used more sparingly. "What drugs will not cure, the knife will; what the knife will not cure, the cautery will; what the cautery will not cure should be considered incurable."[41]

Despite the attention that medical writers gave to both theory and practice, the therapeutic resources of secular medicine were slender. It could mend broken bones, reduce dislocations, cauterize wounds, perform various kinds of surgical operations, engage in venesection or phlebotomy, administer traditional drugs and remedies, and prescribe rest and a regimen that involved change of diet, exercise, and baths.

Geoffrey Lloyd has observed that the theories proposed by many of the writers of Hippocratic treatises are not very convincing to us and their therapeutic procedures are not always impressive either. Not only are these treatises highly speculative, but they contain the widest disagreements regarding the theories proposed. While the approach of the Hippocratic Corpus relies on a methodological naturalism,[42] Lloyd argues that Greek medicine owed its success to a considerable degree to a "gap between theory and practice," which allowed physicians to be cautious in their diagnosis and, sometimes ignoring their own theories, to exercise common sense in devising therapies for their patients.[43] "They relied on what worked."[44] Moreover, medical practice based on a theoretical model often coexisted with elements of folk medicine and magic, hermeticism, and astrology.[45] In fact, a good deal of Greek natural philosophy consists of popular beliefs that were incorporated into authoritative texts, sometimes (though by no means always) with an attempt to provide a rational basis.[46] Medicine was above all a practical subject, and it relied heavily on experience in therapeutics, including folk wisdom, which was seen most notably in drug lore.[47] Thus Dioscorides, an army doctor who lived in the first century A.D., made use of folklore in his collection of drugs for *De materia medica.* Other medical writers incorporated magic and astrology. Soranus (fl. A.D. 98–138), a prolific medical writer who penned far more than the *Gynecology* for which he is best known, was willing to recommend amulets and harmless folk practices if they improved the patient's outlook.[48] Not only was folk medicine absorbed into learned medicine in a manner that would be regarded as unscientific today, but magic and astrology were subjects of serious study by the educated classes in classical antiquity and accepted as valid branches of natural philosophy. The definition of natural philosophy was broader in the classical world than that of science today, and the boundary between theoretical/speculative medicine and the occult sciences was less well defined than it is in the modern world. Medical practitioners could and did differ *toto caelo* in diagnostics and therapeutics, as well as in what was theoretically and medically acceptable, without losing their public recognition as physicians.

Aline Rousselle contrasts two very different fourth-century medical writers from Gaul: Oribasius, who compiled a learned work, *Collectiones medicae,* written in Greek, which represented a late flowering of the classical medical tradition; and Marcellus, who compiled *De medicamentis liber,* a pharmacopoeia that incorporated magical practices. Marcellus's compilation drew on folk medicine taken over from his older contemporary Ausonius. Their approaches to frenzy and melancholia varied according to the etiology assumed by each physician. Oribasius diagnosed mental disorders, such as epilepsy, apoplexy, and melancholy, by their physical symptoms, and he used the same remedies that he would have employed for any other organic disease. Thus he prescribes bleeding and purges, on the assumption that the disorders have natural causes. Our source is silent regarding whether Marcellus believed that some divine or demonic power outside the affected person caused the medical condition. Deducing Marcellus's etiology from his therapeutics, Rousselle infers that he assumed a demonic causation for which he prescribed treatment by a combination of drugs and magic.[49] Formulas, prayers, and magical rites, however, were used in antiquity for ordinary diseases and not merely for demonically induced ones.[50] Whether Rousselle is correct or not in speculating that Marcellus believed in demonic causation of disease, there was no question whether a physician should be consulted regarding madness, only a question of the treatment that the physician might administer. We can identify two physicians in fourth-century Gaul who administered very different kinds of therapy.

Although physicians were widely available during the Roman imperial period, their social status varied enormously from highly educated physicians, whose patients were wealthy, to slaves. They were often (but not always) associated with one or another medical sect and were dependent in a society without medical licensure on a reputation that was based on their success in treating those who sought their help.[51] They had been regarded as sufficiently valuable to society to be granted financial benefits and civic honors by Greek cities in the Hellenistic period to encourage them establish or retain residence. In the first century B.C. Julius Caesar granted Roman citizenship to free physicians who lived at Rome, and this act was followed by the grant of tax immunity.[52] From this time on it becomes difficult to distinguish between Greek and Roman medicine, and perhaps one should speak instead of a continuum whose complexities indicate more than merely a Roman assimilation of Greek ideas.[53] Medicine in the Western Empire continued to be largely practiced by Greek physicians. "Almost 90% of doctors in the 1st cent. AD, 75% in the 2nd, and 66% in the 3rd, are from the Greek East."[54] Most physicians who practiced in the West were either freedmen or slaves.[55] It was not easy to determine the effects of any treatment, and physicians often disagreed with one another in what was a highly competitive marketplace.[56] When confronted with chronic disorders that were painful but not fatal, physicians could do little, a common theme in the Hippocratic Corpus.[57] For this reason they usually declined to treat patients whose cases they considered hopeless, fearing damage to the reputation on which their practice rested.[58] Many sick persons treated themselves for common ailments that they or a family member could alleviate by employing traditional remedies, or they relied on dreams for guidance.[59] Yet despite all the resources available in the classical world, it has been suggested that physicians' ability to heal had not advanced appreciably from the fourth century B.C. to the second century after Christ.[60]

Hebrew and Jewish Medicine

Traditional Hebrew Approaches to Disease

The early Hebrews, like the Romans, were unusual in apparently having no native medical tradition.[61] Like the Romans, they had recourse to folk remedies. The Old Testament contains incidental references to the existence of binders of wounds (Isa. 3:7), knowledge of setting fractures (Ezek. 30:21), and the employment of therapeutic substances (Isa. 1:6; Jer. 51:8, 8:22). But we have no evidence that any systematized therapeutics existed in early Israel. The Israelites appear to have had no practitioners similar to those that existed elsewhere in the ancient Near East, where magic, religion, and medical empiricism easily coexisted in a medical context. The nomadic origins of the Hebrews may in part account for this absence, but a more likely reason is the Israelites' reluctance to employ the magical or pagan healing practices that were found in neighboring cultures, such as those of Egypt and Mesopotamia, for fear of religious syncretism. For the Hebrews, Yahweh was the only healer (Ex. 15:26; cf. Deut. 32:39). There is little evidence, however, to support the widespread view that the Hebrews held generally negative views of physicians or medicine. It is likely that the Hebrews sometimes employed medical craftsmen from Egypt and the Fertile Crescent but forbade the use of pagan religious and magical methods in Israel (Ex. 22:18; Lev. 19:26, 31, 20:6, 27; and Deut. 18:10–11). Yet in spite of this prohibition, some Hebrews adopted magical practices from the indigenous Canaanite population (see Isa. 3:2–3; 2 Chron. 33:6; Ezek. 13:18–20), and Jewish magic later became famous in antiquity.[62]

The Old Testament's incidental references to disease are often related to moral or spiritual factors. Disease was an affliction that was sent by Yahweh, sometimes as a punishment for sin (Ex. 12:12; 1 Sam. 5:6; 2 Chron. 26:20) but also as an act of his will (Hab. 3:5). To say that it was Yahweh who healed did not thereby rule out some perception of natural factors as a cause of disease, even if those factors were undefined or poorly understood. The Hebrews, like their neighbors, conflated natural and divine causation. When the Philistines captured the Ark from the Israelites, they suffered from an epidemic of painful tumors (1 Sam. 5:6–12). But while they recognized the disease as divine punishment, they also saw a connection with the rats that were plentiful in Ashdod and Gath, where the Ark rested. When they returned the Ark to the Israelites, they sent an offering of five golden images of tumors and five of rats (1 Sam. 6:2–5). The episode indicates that the Philistines could posit a causal relationship between rats and disease while at the same time attributing the plague to the anger of Yahweh. Nor did belief in divine agency preclude the employment of natural means of healing, which were known and applied (e.g., 2 Kings 20:7). The occasional appeal to prayer, repentance, and fasting for healing (e.g., Ps. 32:3–5, 38:1–11; Isa. 38:1–6; 2 Sam. 12:16–23) does not imply that they were regarded as typical means of healing. There is no evidence that Israelites of the Old Testament period regarded demons as the cause of ordinary disease. Demonology arose in late pre-Christian Judaism, but we do not find it in the Old Testament. Although the silence of our sources must not be taken as evidence that the Hebrews lacked all medical knowledge, it does suggest that their understanding of the causes of disease remained rudimentary and that treatment was largely confined to folk remedies.

Hellenistic Jewish Medicine

It is not until the Hellenistic period that we have clear evidence of professional physicians within Jewish society. During the Diaspora Jewish communities were exposed to the Greek theoretical understanding of disease and adopted it as compatible with their religious traditions. Greek medicine had been sufficiently divorced from its pagan religious background to be adapted to a variety of belief systems, including Judaism, as a value-free approach to healing.[63] The readiness of Hellenistic Jews to accept Greek medicine can be seen in the Wisdom of Jesus ben Sira, which was written in Hebrew in the early second century B.C. and later translated into Greek in Alexandria by the grandson of the author. In a well-known passage ben Sira urges the reader to honor the physician because God has appointed him to heal.[64] He receives his wisdom directly from God, who also produces medicines from the earth that men ought to employ. To ben Sira healing by physicians is fully compatible with prayer, for it is ultimately God who heals. But the physician seeks the help of God so that his diagnoses will be successful and his treatment will save lives. There is nothing in what ben Sira says about physicians that is discordant with the earlier spirit of Hebrew thought. It is still God who heals, but he does so through the physician, who is his agent.[65] Not all Jews accepted Greek medicine so readily. Philo (c. 13 B.C.–A.D. 45) was more ambivalent in his view of physicians and medicine, as was Josephus (c. A.D. 37–100), although both generally depict them in a favorable light.[66] We know of Jewish physicians who practiced medicine in Palestine during Jesus's time.[67] The Talmud indicates that every city and large place had doctors.[68] In Jerusalem a temple physician was maintained to treat the temple priests.[69]

The Christian Reception of Medicine in the Second Century

Healing by natural means had become part of the general cultural framework in which Christianity arose and spread. Because medicine was, for those who could afford it, the most widely accepted means of healing, Christians were required to define their attitude toward it. The fact that the naturalistic basis of medicine was value-neutral made it relatively benign in religious terms. Its accepted place in the curriculum of general education was a testimony to its cultural authority. While early Christians seem for the most part to have accepted Greek medicine for the healing of disease, one finds nuances in the manner of its reception, as we shall see by examining three important Christian apologists of the second century, Tertullian, Origen, and Tatian, all of whom have been cited as being opposed to medicine.[70] Although none of the three can be said to be hostile to medicine, the approach taken by each was somewhat different.[71]

The second-century apologists were theologians who defended their faith philosophically against the pagan critics of Christianity. It was they who began the process of harmonizing Christian revelation with Graeco-Roman philosophy.[72] Apologists like Justin Martyr (105–167) and Tertullian (c. 160–c. 225) were intellectuals who had received their education in classical culture and were greatly influenced by it. Their opinion of natural philosophy, which was a component of Greek philosophy and inseparable from it, was ambivalent.[73] While they often denounced it for its pagan religious or anti-Christian elements, they employed its methods, thought forms, and vocabulary and drew on it when it was useful in the rational defense of Christianity.[74] These fathers hellenized Christianity by taking over elements of classical culture and incorporating them into a Christian worldview, a process that they referred to as "despoiling the Egyptians" (see Ex. 12:36).[75] There existed, they believed, truth in pagan philosophies, which could be used to advantage by Christian apologists. The earliest apologist, Justin Martyr, writes that "there seem to be seeds of truth among all men."[76] "Whatever things were rightly said among all men," he asserts, "are the property of us Christians."[77]

An early Christian apologist who has been cited as a prime example of Christian anti-intellectualism is Tertullian. Tertullian seems to be just the kind of theologian who could be expected to take an uncompromising stand against both the value of natural philosophy and Christians' use of medicine for healing. He is well known for having argued that, in the relationship of faith to reason and of Christianity to culture, an absolute breach existed between science and faith. His rejection of philosophy was, at least formally, harsh and absolute. "Quid ergo Athenae Hierosolymis? Quid academiae et ecclesia?"

> What indeed does Athens have to do with Jerusalem? What concord is there between the Academy and the church? What between heretics and Christians? . . . Away with all attempts to produce a mottled Christianity of Stoic, Platonic, and dialectic composition! We want no curious disputation after possessing Christ Jesus, no inquisition after enjoying the gospel! With our faith, we desire no further belief. For once we believe this, there is nothing else that we ought to believe.[78]

Yet for all his rhetorical excess in denouncing philosophy (and Tertullian was a formidable controversialist), he was himself a man of wide culture and learning who drew deeply from the well of classical culture at which he had been nurtured.[79] And nowhere is this more true than in the field of medicine. Even a cursory reading of his works will demonstrate that Tertullian knew enough about medicine to make frequent use of medical concepts.[80] It is true that in the *De anima* (*On the Soul*) he condemns physicians who are said to have practiced vivisection (ch. 10) and censures the practice of embryotomy (ch. 25).[81] These references to the misuse of medicine are the most commonly cited evidence in support of his alleged hostility to the medical art. Nowhere in the *De anima,* however, does he condemn physicians or medicine in universal terms. Moreover, it is clear to any reader of his works that Tertullian had studied medicine a good deal, probably as a part of his general education.[82] He makes frequent use of medical analogies to illustrate theological and religious concepts. And he cites approvingly a number of medical writers, most prominently Soranus,[83] who assumed an authority for Tertullian in medical matters that Galen was to have for later Christian writers. The fact that Tertullian cites him frequently and respectfully is difficult to harmonize with his alleged antipathy towards medicine.

More important, Tertullian's works suggest that he had a high regard for both medicine and physicians. Medicine was for him a gift of God. He writes: "Let Aesculapius have been the first who sought and discovered cures. Isaiah mentions that he ordered medicine for Hezekiah when he was sick. Paul also knew that a little wine was good for the stomach."[84] In a number of passages scattered throughout Tertullian's extensive writings, one finds repeated words of praise for the healing art. In chapter 5 of the *Scorpiace,* for example, he points out that the pains administered by the physician are often necessary to produce healing. People foolishly flee a physician who must cut or burn, but such pain is required to produce a good end. And once a man is cured he will praise the physician's skill. This theme was a familiar one among classical philosophers. The whole passage defends martyrdom, which brings about eternal salvation, on the analogy of pain inflicted by a physician that brings about healing. Far from speaking of medicine disparagingly, Tertullian treats it as beneficial to humankind. In an instructive passage in the *De anima* (ch. 2), he contrasts philosophy unfavorably with medicine. He rejects the traditional idea that "medicine is the sister of philosophy." He finds in the particularity of the former a better method than the speculation that characterizes philosophy. Medicine reaches conclusions that are based on data gathered from observation, while philosophy bends its opinions to the laws of nature.[85]

A somewhat different approach is that of Origen (c. 185–c. 254). Origen's view of medicine is a highly positive one.[86] In asking if all knowledge comes from God, he wonders what knowledge could have greater likelihood of divine origin than medicine, which is, after all, merely the understanding of health.[87] God, he writes, provided medical knowledge for humankind, just as he provided herbs and other healing properties.[88] Origen considers medicine "beneficial to mankind" and recommends that Christians employ it for healing.[89] In discussing the use of medicine for healing, however, he distinguishes between two classes of Christians. "A man," he writes, "ought to use medical means to heal his body if he aims to live in the simple and ordinary way. If he wishes to live in a way superior to that of the multitude, he should do this by devotion to the supreme God and by praying to Him."[90] The distinction between the ordinary and the superior Christian is one that is found in many of Origen's writings. It reflects his belief that everything has a double aspect: that which is sensible and can be known to everyone, and that which is spiritual and can be known only to those who are superior.[91] He believes that the more spiritual Christian should rely solely on God for healing and avoid the use of physicians and medicines. Origen is not here introducing a dualistic element into the employment of medicine by Christians. He is, rather, enunciating a principle that came to be adopted by a number of later Christian writers and ascetics, namely, that medicine, however lawful (and indeed efficacious) for all Christians, might freely be renounced by those who seek a closer dependence on God and who look for bodily healing through prayer alone. While this remained a minority view among early Christians, there were always some who maintained it.[92]

Two additional second-century Christian writers who have been regarded as hostile to medicine are Marcion (d. c. A.D. 154) and Tatian (second century). In Marcion's case the only grounds adduced as evidence have been his deletion of the designation of Luke as "the beloved physician" (*ho iatros ho agapētos*) from Colossians 4:14 and his radical dualism. Darrel Amundsen rightly called the first alleged ground "flimsy," while the second is simply insufficient to support the conclusion that Marcion thought medicine inappropriate for Christians.[93] It has been alleged that Tatian opposed altogether the use of medicine for the treatment of illness, but in fact he rejected only the use of drugs (*pharmaka*), which he believed allowed demons to gain entry to the body.[94] Hence, while he had specific reservations, his objections did not amount to a blanket condemnation of medicine.

Tertullian, Origen, Marcion, and Tatian were not in full agreement regarding the place of medicine in the life of a Christian, and indeed we shouldn't expect them to be. Although the evidence regarding Marcion's position is inconclusive, the grounds for asserting that he opposed the use of medicine are too insubstantial to bear the weight that has been placed on them. But none of the remaining three rejects the use of medicine; all accept it with qualifications. The evidence from the second century, however, permits us to go farther in reconstructing the views of mainstream Christians. Positive attitudes to medicine are suggested by the frequency with which Christian writers use medical theories and terminology as analogies for their religious beliefs. Christians studied medicine, which was widely held to be an essential subject for persons of education and culture, as part of their general education.[95] Like Tertullian, Clement of Alexandria (c. 160–215) exhibits a considerable knowledge of anatomy and physiology, which he uses to good effect in illustrating theological themes.[96] "For Origen," writes Bostock, "the art of medicine was the clearest possible parable of the Gospel in action."[97] Thus the medical model of health as the harmony of elements like the hot and the cold or the dry and the wet finds a parallel in the relationship between the body and the soul.[98] Just as physicians apply cautery and bitter medicines to the body, so God applies harsh disciplines to the soul.[99] The medical doctrine of contraries supplies a parallel for the removal of sin and evil.[100] In drawing each of these analogies from medical theory, Origen displays a familiarity with the Hippocratic treatises and the works of Galen. The author of the pseudepigraphical work *On the Resurrection of the Body*, which is attributed to the second-century writer Athenagoras, similarly follows a medical writer, perhaps Galen, in some detail in his lengthy discussion of digestion.[101] This interest in medical theory, which assumes a medical understanding of disease on the part of early Christian writers, continued throughout late antiquity, especially in dealing with theological ideas that have implications for the body, such as resurrection.[102] The Cappadocian fathers, for example, reveal an understanding of theories of contagion in discussing the cause of leprosy.[103] It appears, moreover, that both Basil (c. A.D. 330–79) and Gregory of Nazianzus (A.D. 329–89) studied clinical medicine.[104] John Cavarnos argues that in discussing the relation of the body and the soul, no Christian writer reveals a greater understanding of the physiological basis of sensation than does Gregory of Nyssa, which he derived from his extensive reading in medical sources.[105] Indeed, his works abound in medical terminology, which reflect his keen scientific interest in medical theory.[106] Nemesius (fl. c. 390), bishop of Emesa in Syria, composed a work, *On the Nature of Man*, that was a Christian anthropology that relied heavily on a close study of Greek medicine. Like most Christian writers who appeal to medical theory and practice, Nemesius studied medicine as a branch of philosophy rather than as a physician in training.[107]

Medicine as an Analogy for the Healing of the Soul

One finds among the apologists no break but rather a continuity with classical culture in their appreciation of secular medicine. That continuity underlies the large number of passages scattered throughout Tertullian's writings in which he makes use of body-soul analogies. Here the well-known Christian critic of pagan cultural values reveals his indebtedness to the long tradition in Greek philosophy of comparing the healing of the body to the cure of the soul.[108] Beginning in the fourth century B.C., medical terminology was appropriated for the discussion of ethics.[109] The Greek view of health (*hugieia*) as a state of the body in which all humors operate in harmony provided an analogy for the soul in which moral virtue (*arete*) was defined as a balance of the elements of the soul (e.g., by Plato).[110] Philosophers spoke of the soul as sick or diseased.[111] They called emotions *pathē*, "sufferings." Moderation provided the key to the soul's harmony. "Pleasure ought to be roused in moderation, otherwise we lapse into sickness," says the physician Eryximachus in Plato's *Symposium* (187E). A healthy individual avoided disease and sickness by practicing moderation and self-control (*sophrosune*, which originally meant "soundness of mind").[112] Overeating or overindulgence of the passions not only led to bad health but also created an unhealthy disposition of the soul. Hence medicine and philosophy complemented each other by together enabling one to lead a harmonious life whose end result was happiness. The body-soul analogy was used by writers of nearly all philosophical schools. Epictetus (c. A.D. 55–c. 135) compared the philosopher's school to a physician's consulting room, in which the philosopher deals therapeutically with deep-seated disorders of the mind that are comparable to physical afflictions, whose cure, like surgery, is a painful process.

> The philosopher's school, sirs, is a physician's consulting-room. You must leave it in pain, not in pleasure; for you come to it in disorder, one with a shoulder put out, another with an ulcer, another with fistula, another with headache. And then you would have me sit here and utter fine little thoughts and phrases, that you may leave me with praise on your lips, and carrying away, one his shoulder, one his head, one his ulcer, one his fistula, exactly in the state he brought them to me. Is it

> for this you say that young men are to go abroad and leave their parents and friends and kinsmen and property, that they may say, "Ye gods!" to you when you deliver your phrases? Was this what Socrates did, or Zeno, or Cleanthes?[113]

Christian writers from the second to the fifth century were indebted to the Greek metaphorical use of medical terminology, particularly in the theme of "Christus medicus" (Christ the Physician), which became a popular title accorded to Jesus in the early church.[114] The ideal physician and the physician as an ideal are types encountered with frequency in classical literature.[115] The word *iatros*, "physician," when used figuratively, is not a neutral term. Unless it is modified by a pejorative adjective, it usually carries with it the metaphorical force of a compassionate, objective, unselfish man who is dedicated to his responsibilities. Thus the good ruler, legislator, or statesman is sometimes called the physician of the state. Essentially, it was thought, the statesman is (or should be) to the state what the physician is to his patient. We find this symbolism already in nonmedical literature of the fifth century B.C. Similarly, ancient philosophers were frequently described as physicians of the soul. Regardless of whether the "medicine" they administered was soothing or painful, it was the good of their patients that was always their proper object. The Greeks themselves came to expect much of medicine, and of the physician too. The art of medicine itself carried with it a humanitarian expectation quite apart from any externally defined ideal like that found, for example, in the deontological literature of the Hippocratic Corpus.

The theme of "Christus medicus" is a familiar one, appearing very early (e.g., in Ignatius's Epistle to the Ephesians 7.2, which can be dated c. 117). One finds it employed throughout the second century, and it quickly became a commonplace.[116] It is primarily in its metaphorical sense, and rarely in its literal meaning, that Christian writers describe Jesus as the healer of humankind. Jesus himself had used the metaphor: "Those who are well have no need of a physician, but those are sick; I have come to call not the righteous but sinners" (Mk. 2:17 = Mt. 9:12 = Lk. 5:31). He became the Great Physician, who as the Savior of humankind offers not physical cures but spiritual healing for sin-sick souls.[117] Hence Eusebius, in describing Jesus, quotes from the Hippocratic work *On Breaths*: "A devoted physician, to save the lives of the sick, sees the horrible danger yet touches the infected place, and in treating another man's troubles brings suffering on himself."[118] Eusebius applies these words not to Jesus's healing of the sick but rather to his sufferings and his taking "upon himself the retribution for our sins."[119]

Jesus came in early Christian literature to assume qualities of the ideal physician who unselfishly succors the ill, qualities that were associated with both Hippocrates and Asclepius. Emma and Ludwig Edelstein have emphasized the importance of Asclepius as a competitor to Jesus.[120] Yet in Christian apologetics Jesus became not an alternative healer whose miracles of healing could compete with those of Asclepius but rather the healer of sinners. Thus Clement writes that "our Educator, therefore, is the Word, who heals the unnatural passions of our soul with his exhortations. For quite properly the relief of the diseases of the body is called the healing art and is learned by human wisdom. But the paternal Word is the only Paeonian physician of human infirmities, and the holy enchanter of the sick soul."[121]

Arnobius of Sicca

The early fourth-century rhetorician Arnobius of Sicca has been singled out as one of several early Christian theologians who was opposed to the use of medicine on theological grounds. Vivian Nutton includes him among those, like Origen, who could "argue that the medicine of the physicians was suitable for the average Christian; but for those of higher capabilities . . . prayer and faith alone sufficed."[122] In referring to Arnobius's alleged hostility to physicians, Nutton cites a passage from his *Adversus nationes.* In this passage Arnobius, who had recently converted to Christianity, describes the physician as "a creature born of earth, not trusting to the truth of science" (1.48). Arnobius asserts that in their diagnosis and treatment physicians are often unreliable. This statement has led Darrel Amundsen to a somewhat different view. Amundsen sees Arnobius not as a Christian who opposed the use of medicine but rather as an example of a well-known type who is found in every age, whose alleged bias "transcends religious and other barriers." His is simply a case of that prejudice against doctors that was articulated occasionally by members of all classes of Graeco-Roman society.[123] In fact, however, as I shall make clear below, Arnobius never attacks the medical profession as such but rather the pagan gods who are credited with having effected healing through medicine.

We know little about Arnobius.[124] Nearly all our meager information about him comes from six brief passages that are found scattered throughout the works of Jerome (c. 347–419/20).[125] Arnobius was a distinguished pagan teacher of rhetoric of the early fourth century. He lived in the North African town of Sicca Veneria, which is situated in present-day Tunisia (now Le Kef, southwest of Dougga). He was a lecturer of distinction and vitriolic in his attacks on Christianity. The reign of the emperor Diocletian (284–305), during which he lived, saw the last and most severe of the Roman imperial persecutions of Christianity, which lasted from 303 to 311. Impressed by the examples of Christian martyrdom that he witnessed, Arnobius was converted to the new faith, perhaps in his sixties, after a dream that warned him to turn to

Christ. He presented himself for baptism to the bishop of Sicca. But the bishop feared that his conversion was insincere, and he requested as a condition of his good faith that the learned controversialist compose an apologetic work in defense of Christianity, which he had so long attacked. The result was his only surviving work, the *Adversus nationes* (*The Case against the Pagans*).[126] Following its completion Arnobius was baptized. A late source, Trithemius (1462–1516), says that he later became a priest. We know hardly anything more about Arnobius than this.[127]

It is against this background that we turn to examining the manner in which Arnobius describes medicine and physicians.[128] He groups medicine with "philosophy, music, and all the other arts" as one of the elements "upon which life is built and refined" (2.69.9). Medicine ensures strength, health, and safety.[129] Arnobius includes physicians among converts like orators, critics, rhetoricians, lawyers, and philosophers, who have been "endowed with great ability" but who have abandoned their former beliefs for Christianity (2.5.16).[130] In a series of rhetorical questions he asks, "Do you entrust your bodies' ailments to the hands of physicians, not believing that the diseases can be relieved by the lessening of their severity?" (2.8.15). And he concludes book 1 by employing the familiar Christian theme of "Christus medicus," Jesus as the Great Physician. Arnobius likens Jesus to the physician who comes from afar, offering medicine that will prevent every kind of disease and sickness (1.65). "Even if the matter were doubtful," he writes, "you would yet put yourselves in his care and you would not hesitate to drink down the unknown dose, induced to do so by the prospect set before you of gaining health and by a love of security" (1.65.4). While Arnobius employs the figure of the physician metaphorically to refer to Jesus as the healer of the afflictions of the soul, its use suggests his high regard for practitioners of the medical art.

The only instances in which Arnobius writes in less than the highest terms of physicians are those found in the context of his discussion of Aesculapius's ability to heal (Aesculapius was the Roman form of Asclepius). He devotes a good deal of space to challenging common belief in the pagan gods, particularly in their role as patrons of the respective arts. In book 3 he tries to undercut the claim that the gods protect human activities by pointing out that they fail to preserve life and health as often as they seem to grant it. Thus, in spite of the belief that Aesculapius presides over medicine, the fact remains that there are many who are *not* restored to health by medical treatment, "whereas, instead, at the hands of those who care for them," he writes, "they even become worse" (3.23.24). Arnobius intends no denigration of physicians here. Indeed, similar sentiments can be found scattered throughout ancient medical literature, reflecting the uncertainty that was recognized to accompany any kind of medical treatment, in which it was possible for the physician to harm as well as to help. The failure, Arnobius suggests, belonged to Aesculapius. His mention of the healing god is usually confined to a passing reference in lists of pagan deities, but in one passage in book 7 (chs. 44–48) he discusses him at length. He challenges the claim that Aesculapius had kept the Romans free from plague, asserting instead that the god had failed to heal many plagues that had arisen in Rome after his arrival there from Epidauros. Most of his argument, however, is devoted to pointing out the incongruity to be found in the legend that the venerable god of healing took the form of a lowly serpent.

But in the most telling passage that he directs against the claim of Aesculapius to heal, Arnobius replies to those pagans who contrast the numerous healing miracles of Aesculapius with the small number that are attributed to Christ (1.49). In a poignant passage Arnobius strikes at the weakest chink in the armor of Aesculapius's propagandists: the failure of many pilgrims to receive the healing they sought from the god. How many thousands of those who went to the temples, he asks, prostrated themselves before the god, offered prayers, and made vows but were *still* not healed? "What good is it, then," he writes, "to show that one or another was possibly cured when to so many thousands no helper has come and all the shrines are full of the wretched and the unfortunate?" Arnobius offers the distinctively Christian response to the claims made for Aesculapius. To those who fail to obtain healing from the god, Christ offers his compassion. He offers it to all who suffer, to the deserving and the undeserving, to good and bad alike. And he turns away no one.

We must now return to Arnobius's alleged denunciation of physicians in the quotation with which I began. Arnobius describes the physician as "a creature born of earth, not trusting to the truth of science" (1.48). The passage is found in the context of a long discussion in which Arnobius seeks to demonstrate the divinity of Christ by appealing to his miracles (1.43–56). It is Christ alone, he writes, who heals the illnesses that have been brought on the human race by the decrees of fate. He thereby shows himself to be even more powerful than the Fates themselves.[131] But, responds an imaginary pagan opponent, other gods have also healed the sick by giving them remedies. Arnobius replies that while that may be true, they have not healed without medicine. Christ required only a simple command or a mere touch to heal; in contrast, the gods heal through natural means. In some cases they prescribe medicine; in others, certain foods or a draught of some kind, or a poultice, or rest and physical exercise.

Arnobius lived long after the change that had taken place in the cult of Asclepius during or before the second century. Until then those who sought healing through incubation in his temples believed themselves to be healed miraculously. By the second century pilgrims, at least at Pergamum, often received detailed instructions from the god in a dream for pursuing a particular regimen or a medical remedy for their illness. "Many of the therapies ordered by the god at the Asclepieion of Epidaurus," writes Vivian Nutton, "differed only slightly from those of the physicians."[132] This kind of healing, thought Arnobius, deserves no great admiration, if one examines it seriously. "[Y]ou will discover that physicians heal in this same way, a creature born of earth, not trusting to the truth of science, but employing the art of guessing and wavering in conjecturing possibilities." His point is not that medicine is unreliable and therefore to be condemned; it is, rather, that medicine is a conjectural art that is unpredictable in its outcome. There is no special merit (i.e., nothing miraculous) in removing illness by medical means, he writes. The healing qualities belong to the drugs, not to the healers. And though it is praiseworthy to know which drugs or treatments are suitable for healing, the credit for knowing this ought to be assigned to the man who heals, not to the god. There is nothing disgraceful about the physician who improves someone's health by the application of external means. The physician is, after all, "a creature born of earth," that is, he is not divine. What *is* disgraceful is that the divine Aesculapius is not able to effect healing apart from the aid of external means.[133]

Arnobius's complaint, then, is not against physicians, who do their best to heal disease by means of the remedies available to them. He is attacking Aesculapius, who is renowned for his claim to offer miraculous healing but who instead merely employs medical means. The god heals as a physician does, by natural means, with all the uncertainties and attendant difficulties of ordinary medical treatment. What he performs may be healing, but it is not *miraculous* healing. In fact, writes Arnobius, many who seek his aid are not healed at all. In describing the physician as one who is "not trusting to the truth of science," Arnobius does not thereby deprecate medicine. His tone here reflects his general pessimism about achieving certainty of knowledge in any sphere of endeavor. The extent of human ignorance and the difficulties inherent in all fields of investigation are common themes in Arnobius.[134] The art of medicine is no exception. Like all the arts, it has an uncertain outcome.

Arnobius's opinion of medicine is of particular interest because the *Adversus nationes* is the work of a recent convert to Christianity, who was anxious to assail the false beliefs of his former religion. If the place of medicine in physical healing were an issue of dispute between pagans and Christians, we should surely find the matter addressed here. If some Christians condemned medicine as incompatible with dependence on the God who heals our physical ailments, Arnobius might be expected to espouse that point of view. Yet not only was the issue not raised, but Arnobius was respectful of both those who healed and the means by which they practiced their art. One finds no denunciation of physicians in Arnobius because Greek medicine had been accepted by Christians generally as fully compatible with Christian theism. The real issue was not one of natural versus miraculous healing but which god was the true source of healing. Aesculapius was a potent force in the third century, and he was regarded by Christians as a threat to the true faith.[135] Hence Arnobius's quarrel was with him, not with physicians. In demolishing the basis of pagan claims to miraculous healing, which remained a powerful apologetic for the old religion, it was essential that he demonstrate that Aesculapius's miracles were not really miracles and that he was not the universal healer that his proponents claimed. This was the crux of the issue, and it is the focus of Arnobius's attack.

On the basis of several statements that have been taken out of their context, Arnobius has been placed among a group of early Christian theologians who were allegedly hostile to physicians. I have attempted to demonstrate that a careful reading of his *Adversus nationes* suggests that he neither opposed the use of medicine nor denounced physicians. On the contrary, he displays a high regard for both. Like the second-century apologists who have also been cited as early Christian opponents of medicine, Arnobius proves on closer examination to have accepted medicine as a gift of God and therefore appropriate for the use of Christians.

Conclusion

The reception of Greek secular medicine by a variety of disparate cultures and religious communities made it possible for a wide spectrum of religious believers to accept the natural causation of disease and to think that commonly encountered diseases were susceptible of treatment by natural means. Because secular medicine was not dependent on a polytheistic worldview, it proved compatible with the tenets of many Jews as well as those of most Christians, who without difficulty assimilated it into a Christian framework. A related factor in the ease with which Christians appropriated Greek medicine was its long history of metaphorical usage. Drawing on classical philosophers, the church fathers found in it a congenial means of illustrating theological ideas. Here, as elsewhere, educated Christians could find much in classical culture that was suitable for appropriation and that provided them with links to a world they valued in spite of their sometimes strong protestations to the contrary.

So deeply ingrained in modern scholarship is the belief that the earliest Christians routinely employed miraculous or ritual healing and that they condemned medicine and distrusted physicians that one finds a persistent reluctance to admit that Christians were, from the beginning of the movement, willing to use medicine without scruples. Anne Merideth writes that, in assessing the opinions of the church fathers regarding whether Christianity and medicine were harmonious or hostile, scholars "far too frequently conflate the evidence from sources spanning more than three centuries with little explicit awareness of the massive social, cultural, economic, political, and religious changes which transformed the Roman world from the second to the fifth centuries."[136] Merideth herself takes a middle position, arguing that early Christians distrusted physicians but that as Christianity grew in status and numbers, attitudes to the culture shifted, with more Christians accepting the medical profession and even becoming physicians themselves. A "cultural relocation of Christianity" occurred in the fourth century, a result of which was the full Christian acceptance of medicine, which some earlier Christians had condemned. Merideth is correct when she states that one finds among Christian writers a "spectrum of opinions about when it is acceptable to resort to the use of medical healing," sometimes even within writings of the same author. But it is a spectrum that can be found throughout the entire range of early Christian history. Christian authors writing in the second century were not hostile to medicine (not even Tatian, whom she cites as an example). But some of them did qualify their recommendations regarding its use. Similarly, criticism of physicians is found throughout antiquity and not primarily, as Merideth suggests, in hagiographical texts (where it is often a rhetorical device for showing that holy men could heal difficult cases that physicians could not). Indeed, the theme is a leitmotif in classical literature.

With a few exceptions (e.g., Cato and Pliny the Elder), conventional criticism of physicians probably tells us little about whether either classical or Christian authors accepted or rejected the use of medicine. There is no trajectory from the second to the fourth century that indicates a growing acceptance by Christians of secular medicine. More purported miracles of healing can be found in the fourth century than in the second, but then more Christian physicians can be found too. The increase in miracles indicates not a diminishing opposition to medicine among Christians, but the influence of cultural changes on healing practices, in particular, the rise and influence of the holy man. The increase in the number of Christian physicians in the fourth century is largely attributable to the legalization of Christianity and to the rise of charitable medical institutions in the later half of the century, which Christians viewed as a vehicle for medical beneficence. The evidence suggests that acceptance of medicine by Christians, sometimes with qualifications that had chiefly to do with spiritual and pastoral concerns, coexisted with occasional criticism of physicians, which employed the same charges, in fact, that classical writers made. Occasional criticism or ridicule at their expense had nothing to do with "a larger rejection of Greek culture and learning."[137]

It has become common for medical historians to borrow the language of medical anthropologists in speaking of "a plurality of etiologies [that] corresponds to a plurality of therapies."[138] "In a world such as the eastern Roman Empire in the fourth century," writes Merideth, "in which multiple etiologies were invoked to explain the cause of illness, it is not surprising to find a similar multiplicity of healing options available to those who sought the alleviation of their sufferings."[139] Historians of medicine frequently speak of the variety of medical systems that were available to the sick in antiquity.[140] Peter Brown describes the "medical pluralism" of ancient society, in which no therapeutic system had final authority.[141] Theoretical medicine and religious healing existed side by side as complementary systems. The coexistence of medical, religious, folk, and magical healing traditions has often been found—and can still be found today—in many societies. Brown cites what anthropologist Lola Schwartz has termed "illness behavior," in which the sick person appeals to a "hierarchy of resort." The pattern is predictable and is found across cultures. The person who requires treatment for a medical condition seeks the advice of family and friends. He or she is attracted first to local healers—those readily available in the village, whether physicians, or holy men, or folk healers. Only when failing to find help there does the sick person seek cures from outsiders. The competition among healers of all kinds offers the sick person a multitude of opportunities for therapy.[142] Geoffrey Lloyd identifies five "demarcated groups" of healers in the ancient world: root cutters (Gk., *rhizitomoi*; L., *herbarii*), drug sellers (*pharmakopōlai*), midwives (Gk., *maiai*; L., *obstetrices*), religious healers, and the physicians represented by the Hippocratic texts.[143] Vivian Nutton's list is longer: herb cutters, druggists, midwives, gymnastic trainers (L., *iatraliptae*), diviners, exorcists, and priests.[144] Some—but not all—sick persons sought therapy from different healers because they attributed their illnesses to different etiologies. But different kinds of healers performed different functions.[145]

While the current focus on the medical pluralism of the ancient world provides a necessary corrective to the traditional view, which saw physicians and healers as synonymous, I believe that it has been overemphasized to the degree that it denies that physicians played a predominant role in the healing of disease in the Roman Empire.[146] We must make some distinctions, however, if we are properly to assess their position in the medical marketplace. To begin with, there existed no "medical profession" as we know it today, with education, expertise, and ethical standards of the kind that guild or state regulation imposes.[147] Judgments of moral probity played a much larger role in the public's assessing the status of practitioners than they do today, and there was no recognized way of defining medical orthodoxy or even quackery.[148] Because of the lack of accepted professional qualifications and the importance that medicine played in the literary culture of Rome, the boundary between an educated layman and a physician was far less defined than it is today. In the words of Vivian Nutton, it "had at least as much to do with an individual's perception of his own role and position in society as with his competence in book-knowledge or practical experience."[149] In a society without licensure anyone could "practice" medicine. Hence the spectrum of competence was broad. Some physicians were barely, or not at all, competent by the standards (lay or professional) of the day, and no professional organization or official body existed to separate the quack from the competent professional. In order to receive Roman citizenship under Julius Caesar or his successor Augustus, the physician was required only to appear before a magistrate and affirm that he was a physician. No examination of training or professional qualifications was required.[150] In assessing how many people had access to medical treatment by physicians, we must also make geographical, class, and urban-rural distinctions. Physicians congregated in towns to the disadvantage of rural dwellers.[151] It was urbanization as much as the spread of Greek physicians throughout the Roman Empire that made medical care widely available.[152] Cities like Rome and Constantinople had an abundance of medical specialists.[153] Yet even small towns might boast a surprisingly large number of physicians, as did Metapontum in southern Italy, which had sixteen.[154] The sick who lived on the edges of the empire were less well served in seeking medical treatment. But even the most remote boundary areas attracted physicians, probably most of them serving with the Roman legions, who brought with them the medical traditions of the Greek world, as did a doctor by the name of Antiochus who penned an elegant dedication to Asclepius, Hygeia, and Panakeia in Chester, England.[155] Those who could not find medical help available locally sometimes traveled to towns where it was available or took advantage of itinerant physicians.[156] Self-help manuals existed to give advice to the layperson who wished to maintain a healthy regimen or who had no access to physicians. Some manuals were written by physicians, others not. They included Rufus of Ephesus's *For the Layman*, Celsus's *De medicina*, and Plutarch's *De tuenda sanitate.*

By the second century of the Christian era there was little distinction between Roman and Greek medicine; the real divide was between town and country, between naturalistic approaches to healing practiced in cities and older, native traditions of folk medicine and medical self-help that lingered in rural areas.[157] The poor and slaves, in disproportionately large percentages, would have been unable to afford physicians.[158] Many peasants would have relied on folk remedies of the sort found in Pliny and Cato. It is likely that a large proportion of the population of the Roman Empire, primarily those outside cities, had little access to a physician.

> It must also be kept in mind [writes Aline Rousselle] that not all diseases were treated. Peasants became more bent over year by year. People were crippled by rheumatism, which periodically caused them pain. All of this was common, expected, and considered a normal part of life. People were patient. They had learned from childhood to put up with pain. They felt powerless to deal with a fever that might go away the next day. They also had recourse to remedies handed down through the generations: special diets, wine, herb teas, compresses. Only after all this had been tried, and only in cases of painful and protracted illness, did they seek out or summon the "specialist" of the village or neighborhood, a sorcerer, healer, or expert in plants or charms. In most cases, illness was dealt with only on these three levels.[159]

Many rural areas had their own folk-healing specialists, such as the Marsi, who lived in the highlands of central Italy and whose reputation as snake charmers made them purveyors of remedies and antidotes.[160] But, in spite of the recent attention that folk healers have received, they rarely appear except in little-read literary and epigraphic sources. While this argument from silence might be taken as evidence for their limited appeal, one might ask how important we would have considered Galen in the Roman medical world of the second century after Christ if his works had not survived. Such arguments are not very reliable. The frequency with which physicians appear has been attributed in part to the class values that underlie much Greek and Roman literature.[161] I think it more likely that it is largely due to the preeminence they enjoyed among healers in the Roman Empire. Although scholars have given increased emphasis to the competition of healers in the medical marketplace, Merideth is correct, I suspect, in arguing that the situation was less one of competition than it was of a "hierarchy of resort." Competition, observes Rebecca Flemming, was fiercest not between but within the medical professions.[162] For those who could afford them, physicians were ordinarily the healers from whom the sick sought treatment. While folk healers who offered amulets, herbs, and help from astrology were abundant, home remedies administered within the family or by relatives and friends may have been the most common form of medical therapy. Despite this, the healers who are visible in our sources were not folk healers but physicians.[163] And not in upper-class literature alone. In Mark's Gospel a woman who had suffered from a uterine hemorrhage for twelve years approached Jesus in a crowd, touched his garment, and was healed.[164] The woman told Jesus that she had endured a good deal from many physicians and that she had spent all that she possessed but had grown worse, not better (Mk. 5:25–34). She is specific: she had consulted not a variety of healers but many physicians. Only in desperation did she seek out Jesus, hoping that he could provide the healing that physicians had been unsuccessful in effecting. Although we may speak of a "hierarchy of resort" here, it does not encompass competitive alternative healers but rather a miraculous healer whom a woman with a chronic illness sought out after years of failure to gain healing from *physicians.*

We find the same pattern in later hagiographical literature: it was when the sick could gain no help from physicians that they resorted to holy men.[165] The evidence from the Roman imperial period suggests that physicians were not merely competitors in a marketplace peopled by a variety of healers but that they were themselves the healers whom those in need of medical attention ordinarily consulted first when they had the opportunity and the resources. Healing shrines, like miracle workers, were not the first choice of most (except the poor) who needed medical help but often the last resort for those who had found no help from physicians. Physicians and ascetics sometimes consulted with one another and referred patients who could be better helped.[166] The competence and therapy offered by *medici* and *iatroi* varied from physician to physician. Some were highly trained and had adopted one or another theoretical approach, while others had little or no medical training and would be considered incompetent charlatans both by their own professional colleagues and by a lay public that was often highly literate medically. To say that Christians sought the aid of physicians does not suggest that the treatment they received was necessarily efficacious, only that it represented some form of medical therapy from a self-proclaimed physician. Physicians are likely to have been consulted routinely by Christians. Ritual healing (such as anointing) is unlikely to have been the first resort of Christians who sought healing but a pis aller for the chronically ill or for those who had failed to gain healing by conventional medical means or perhaps by any therapy at all. It is significant that it is not alluded to by second-century Christian writers, who speak highly of doctors and the art of medicine, although of course it is possible that it was present but that they merely chose not to mention it. But in regard to the prevalence of ritual healing, it is more reasonable to think that the silence of our sources reflects the practice of the early Christian communities.

It would be a mistake to conclude that the reception of Greek medicine (or as we can now call it, Graeco-Roman medicine) was a passive process of Christian cultural accommodation or merely a by-product of the hellenization of Christianity. The attraction of Greek philosophy for intellectual Christians led to a good deal of tension that, in the case of the apologists, often involved initial denunciation followed by gradual acceptance of much that was not in direct conflict with Christian theology. But medicine, like natural philosophy, could be detached from its pagan framework with relative ease. Its functional separation from Greek mythopoeic structures and its focus on secondary causes permitted its appreciation by Christians, who inhabited a world of overlapping cultures in which secular medicine had become so much a part of ordinary life in the Roman Empire that it was everywhere taken for granted. "We are dealing," writes Vivian Nutton, "with [merely] one segment of a Mediterranean society, to borrow Goitein's description of later Judaism, in which many features of everyday life, including medicine, can be found elsewhere around the shores of the Mediterranean—in Alexandria, Rome, or Ephesus."[167] Christians and their neighbors of other religious or ideological backgrounds sought healing from the same value-neutral medicine.[168] The issues raised in the Christian reception of secular medicine were chiefly theological ones. Thus some Christian writers saw in the very promise of success in the healing art a hidden danger, one that had been recognized much earlier, in a Jewish context, by ben Sira: by crediting to medicine the ability to heal, one threatened to replace dependence on God with reliance on medical means alone. It was a danger that was as old as the Jewish king Asa, who resorted to physicians rather than seeking the guidance of Yahweh (2 Chron. 16:12), and it was frequently condemned by the fathers.[169] The most notable transfer of pagan symbols of compassionate healing was in the substitution of Jesus for Asclepius and Hippocrates as the ideal physician. While Jesus did not fully displace Asclepius, over time he inherited the metaphorical and iconographic status that had become attached to the healing god and, at a later date, some of his healing functions as well.[170] All this was the result of cultural negotiations that were the more easily accomplished because medicine enjoyed nearly universal recognition in the classical world as a humane art that transcended local cultures and particular ideologies.

Early Christian Views of the Etiology of Disease

On opening the pages of the New Testament, many modern readers find themselves in what appears to be an alien world, in which supernatural forces intervene in ordinary life. The Gospels focus on the extraordinary Palestinian ministry of Jesus, who casts out demons and miraculously heals the sick of every description.[1] The Book of Acts recounts the activities of Jesus's apostles, who themselves exercise miraculous healing and exorcism and carry their supernatural gifts throughout the Mediterranean world, a world in which they themselves encounter exorcists and magicians. If we were to describe early Christian beliefs regarding sickness and healing on the basis of a cursory reading of the New Testament (especially of the Gospels), we might be inclined to summarize them as follows. Disease is caused by sin or by demons and is healed supernaturally. Some illnesses can be cured only by exorcism, others by miraculous healing. Still others are susceptible of healing by prayer, faith, or anointing. With slight modification this description can be found in several standard studies of the role of medicine in the early Christian church, including those written by biblical, classical, and medical scholars.[2]

On the assumption that physical impairment is associated with demonic activity in the Gospels, a number of scholars have concluded that early Christians altogether rejected a naturalistic etiology of disease, regarding demons as the cause of all, or at any rate most, illness. Thus Ulrich Mueller writes, "[T]he dominant view of the New Testament is that demons are the causes of sickness."[3] Otto Böcher has provided the most extensive case for this thesis. In a trilogy of studies Böcher claims to find in the New Testament numerous implicit allusions to demonic power and its influence on disease.[4] While this "pan-demonological interpretation," as one historian has called it,[5] is not accepted by all New Testament scholars or historians of early Christianity,[6] it can be said to be the currently dominant narrative. For example, L. D. Hankoff has written a psychohistorical study in which he examines the kinds of miraculous healing employed by Jesus, with specific reference to the role that suggestion and imagination might have played both in the healing itself and in its portrayal in the Gospels.[7] Hankoff compares Jesus's miracles of healing to similar kinds of healing that are recorded in the Jewish Talmud of the early centuries of the Christian era. He suggests that significant disparities existed, arguing that Jesus and the writers of the Gospels differed from the Talmud in attributing all disease to demonic etiology. "In so elaborating the demonic element in spiritual healing," writes Hankoff, "the New Testament authors advanced a disease concept of nearly universal proportions. A single cause could explain all illness."[8]

I propose to examine the early Christian understanding of disease, specifically the thesis that early Christian sources ascribed all illness to demonic etiology. I shall attempt to demonstrate that the evidence drawn from Christian writers of the first five centuries fails to support the pan-demonological view of scholars like Mueller, Böcher, and Hankoff. After sketching the origins and nature of belief in demons in Jewish culture, and the role of demonology in early Christianity, I shall contend that Christians typically accepted a natural causality of disease, which was an inheritance from the Greek understanding of illness. Finally, I shall suggest that in their theological understanding of disease, Christians viewed illness within a conceptual framework that permitted them to assume the existence of secondary (i.e., natural) causes but to explain them in ultimate terms as the manifestation of God's will.

Exorcism in Intertestimental Judaism

Edwin Yamauchi suggests that one finds four models of disease etiology in the ancient world.[9] The first model postulated a deity as the direct cause of retributive disease. The second assumed that supernatural beings other than gods (e.g., demons) inflicted disease on individuals. The third ascribed sickness to magic, which was most often performed by a sorcerer or magician. The fourth presumed natural causes, which could be discovered by experience or investigation. Every ancient society held to one or more of these etiological models of disease. They were not mutually exclusive but were often combined, as they have been in nearly every human culture. Each of them called for a particular treatment. A divine etiology demanded prayer, sacrifice, offerings, or confession. A demonically caused illness called for exorcism or another form of supernatural healing, such as prayer. A magically induced illness required countermagic, while a disease that was attributed to natural causes required medical treatment but was sometimes believed to have been healed miraculously (e.g., by Asclepius or Jesus). Treatment not infrequently consisted of more than one of these therapeutic approaches.

While demonology long played a prominent role in the belief systems of Babylonia and Persia, it is not until the postexilic period, after the resettlement of Palestine by Jews who returned from Babylon in c. 538 B.C., that one sees frequent reference to the activity of demons and exorcism in Jewish literature.[10] Both reflect syncretistic elements in intertestimental or Second Temple Judaism. The apocryphal book of Tobit, written probably in the third century B.C., provides the earliest account of exorcism in Hebrew literature.[11] Tobias the son of Tobit is described in the narrative as having caused the demon Asmodeus to flee by burning the heart and liver of a fish that created an unpleasant odor (Tobit 8:2–3). We know of several Jewish exorcists and magicians, but they cannot always be distinguished from wandering charismatics who healed the sick and performed nature miracles. Among them were Honi the Circle Maker, Yohanan ben Zakkai, Hanina ben Dosa (who is said to have healed by prayer), and Eliezer ben Hyrcanus.[12] The most prominent was the Jewish exorcist Eleazar. According to Josephus, who had seen him perform an exorcism in the presence of the emperor Vespasian, his officers, and his troops, he placed a ring that held a root that Solomon had prescribed near the nostrils of a demoniac. The demon was drawn out and commanded not to return as Eleazar recited the name of Solomon and uttered incantations that were attributed to him.[13] In the intertestimental period Solomon came to be regarded as a celebrated magician and exorcist based on his ostensible knowledge of plants and animals and power over spirits and the virtues of roots (see 1 Kings 5:12 and Wis. 7:15–22). He was said to have composed incantations to cure illnesses as well as forms of exorcism.[14]

The Essenes are known to have practiced exorcism. They were famed for their skill in employing magic, exorcism, and folk medicine for healing.[15] Several Qumran documents attribute legendary afflictions of pagan kings to demons who were expelled by Jewish exorcists.[16] In fact, Jewish exorcists enjoyed a high reputation among pagans throughout the Roman Empire. One finds evidence of Jewish exorcists in the New Testament (Mk. 9:38–40 = Lk. 9:49–50), among them the seven sons of Skevas, a Jewish chief priest (Acts 19:13–16). The magical papyri preserve a number of Jewish incantations and formulas of exorcism.[17] It is difficult to estimate how widely first-century Jews outside sectarian groups like the Essenes and those who sought out Jewish exorcists believed in the demonic etiology of disease. Although that belief has left its mark on rabbinic literature, it does not appear to have been the predominant view of disease among Palestinian Jews, and given its often sensational nature, it is likely to play a more prominent role in the surviving evidence than it played in fact.

Disease Etiology in the New Testament

Underlying the view that early Christians ascribed disease wholly or largely to demons is the assumption that the Gospel accounts of Jesus's exorcisms reflect contemporary Jewish views of demonology. The evidence, however, does not suggest that Jesus shared the demonology of his Palestinian contemporaries. The Gospels do not record either Jesus's explanation of the phenomenon of demonic possession or that of the Evangelists themselves. But although in individual instances some similarities existed between Jesus's methods and those used by Jewish exorcists, in general, the differences are much more significant. Jewish exorcists used magical means, such as amulets, both to prevent and to cure disease, and exorcism, either by outward means or by formulas of incantation, to expel disease-causing demons.[18] The Gospels attribute neither magic nor incantations to Jesus, whose method was, with few exceptions, always the same: he spoke a word, and the demon departed (see, e.g., Mk. 1:23–26 = Lk. 4:33–35). Those who witnessed Jesus's exorcisms were amazed not only by the method he used but by Jesus's authority over demons. Both differed from what they expected (Mk. 1:27–28 = Lk. 4:36–37).

The Gospels portray Jesus's healing miracles as "signs" (*ta sēmeia*) that provided evidence of his messianic mission. For this reason the Gospels distinguish them from the miracles of both magicians and exorcists. The frequency with which demons appear on the pages of the Gospels reflects the Evangelists' belief that the advent of Jesus's kingdom brought about a spiritual conflict with the forces of Satan. Jesus's exorcism of demons was one dimension of this conflict, which they viewed as "a cosmic struggle in history to inaugurate the eschatological reign of God."[19] Mark describes more of Jesus's exorcisms than does any other Gospel writer, and they appear as one of the motifs of Jesus's ministry, especially in the first half of Mark's Gospel (beginning at 1:23–28), where the writer's portrayal of Jesus reflects his theological agenda. Chapters 1 through 8 picture Jesus as a powerful miracle worker, through whom God, the Great Warrior, is undoing the evil brought about by Satan's control of the world. Hence in his frequent confrontations with demons Jesus repeatedly challenges the powers of darkness. Beginning at 8:31 we enter the second half of the Gospel, in which Jesus appears in a very different role, as the suffering servant of Isaiah. In spite of the frequency with which exorcisms appear in Mark, there are several indications that neither Jesus nor the Evangelists believed that disease was ordinarily caused by demons. First, while the Gospel writers speak of Jesus's healing and exorcism as related aspects of his messiahship, they routinely distinguish between them, as Matthew does when he writes, "[A]nd he cast out the spirits with a word, and cured all who were sick."[20] Exorcism and healing denoted different aspects of Jesus's messianic ministry, not a single act. Second, a number of medical conditions are described in the Gospels. They include deafness, muteness, blindness, leprosy, fever, dysentery, a uterine hemorrhage, lameness, paralysis, dropsy, and a withered hand. While no immediate cause is given for any of these illnesses, most of those mentioned in the Gospels (indeed, in the New Testament) fall into the category of ordinary diseases or congenital conditions.[21] Third, the symptoms given for diseases or physical impairments are for the most part distinguished from the symptoms that are said typically to accompany demonic possession, such as erratic or self-destructive behavior (as in Mk. 5:1–5 and Mt. 8:28–29).

Three instances exist in the Gospels in which demonic possession accompanies physical impairment. In the first a dumb man possessed of a demon is brought to Jesus, who expels the demon, after which the man is able to speak (Mt. 9:32–33 = Lk. 11:14). In the second a man who is possessed, as well as blind and dumb, is brought to Jesus, who heals him so that he is able both to speak and to see (Mt. 12:22).[22] In neither case does the person act in the manner usually associated with possession in the Gospels. Both men are physically impaired, however, and while in neither case is the impairment attributed to the demon that Jesus exorcized, the natural inference of the narrative is that it was demonically induced, although that conclusion is not the only one possible.[23] In the third instance a boy described as epileptic or moonstruck (*selēniazetai*) is brought to Jesus. His is the only example in the New Testament of a demon-possessed person who exhibits both self-destructive behavior and symptoms of physical illness (Mt. 17:14–20 = Mk. 9:14–29 = Lk. 9:37–43). These three are the only cases in the Gospels in which a physical illness or dysfunction seems to be attributed by the Evangelists to demons, and it is notable that all are associated with possession rather than with the attribution of demonic etiology.[24] Many more instances exist in which no causality is given and no mention of demons is made.[25] Since the ill are healed without the expulsion of demons, to suggest a demonic etiology in such cases is to read one's interpretation into the text. To say that *no* physical impairment is attributed in the Gospels to a demon flies in the face of the evidence. But one can say that the three instances cited constitute a relatively small number and that they do not suggest that either Jesus or the Evangelists held in general to a demonic etiology of disease.

In neither the classical pagan nor the early Christian world did the attribution of disease to natural causes preclude recourse to miraculous healing.[26] Certainly in the classical world miraculous healing by Asclepius was regarded as efficacious for diseases for which natural causality was assumed.[27] If those who sought healing from Jesus did so because they believed that demons had caused their diseases, they would probably have expected exorcism rather than healing. In most reported instances of illness, however, Jesus is said physically to have healed the sick person rather than to have expelled demons (as in the case of the paralytic in John 5:2–9). It is, in fact, an important distinction of Jesus's treatment of the sick that he *healed* them rather than cast out demons. The conditions that he healed were often congenital or chronic, precisely the kinds of impairments that physicians had not cured. Mark's Gospel records, for example, the case of a woman who had suffered from a uterine hemorrhage for twelve years. She approached Jesus in a crowd, touched his garment, and was healed. Mark was not castigating doctors when he said that "she had endured much under many physicians, and had spent all that she had."[28] The helplessness of physicians (*derelictus a medicis*—"given up by the doctors") was a commonplace in Greek and Latin literature.[29] Hers was a chronic condition, perhaps a nearly hopeless one. But it was an impairment, nevertheless, for which she seems to have assumed natural causes. It neither demanded a demonic explanation nor is given one in Mark's Gospel. As in the case of the man born blind (Jn. 9:1–21), there is no hint of demonic etiology. Of course, there were Palestinian Jews in Jesus's time who did not share a belief that illness was ordinarily to be explained by natural causality. The existence of exorcists indicates that some Jews ascribed at least some illness to demons.

The Book of Acts describes the spread of Christianity throughout the Mediterranean world by Jesus's disciples (the apostles) and Paul. While the apostles are reported in Acts to have performed miracles of healing,[30] the writer of Acts continues to distinguish between healing the sick and exorcizing demons.[31] The frequency of reported miracles in Acts, however, is considerably smaller than is that of miracles in the Gospels. Acts records only four instances of the apostles confronting demonic possession. In three instances the disciples are said to have *both* healed the sick *and* cast out demons.[32] In two of these instances (Acts 5:15–16 and 19:12) generic words are used (*astheneis*, "the sick," and *tas nosous*, "diseases"). In the third (Acts 8:6–7) Philip is said not only to have cast out demons but also to have healed those who were paralyzed and lame *(paralelumenoi kai chōloi)*. In a fourth case Paul is said to have cast a spirit of divination out of a slave girl at Philippi, but no illness or physical impairment is alluded to (Acts 16:16–18). It is noteworthy that we find mention in Acts of exorcists who attempted but failed to cast out demons in Jesus's name. The recorded incident suggests that attempts to exploit the power inherent in Jesus's name merely for purposes of magic or exorcism were unsuccessful (Acts 19:13–17).[33]

In no case, however, are the disciples said in Acts to have cured disease or physical impairment by exorcism. In fact, only a small number of miraculous healings are recorded in Acts.[34] They include healing the lame (Acts 3:1–11 and 14:8–10), a paralytic (Acts 9:33–34), and a man suffering from dysentery (Acts 28:8). All appear to have had a natural causality, none is attributed to demons, and exorcism is not employed in any instance.

When we turn from the Gospels and Acts to the New Testament Epistles and the Apocalypse, we find that demonic possession is notable only for its absence.[35] In fact, in the Pauline Corpus the word "demons" (*daimonia*) is used in only two contexts, neither of them associated with sickness or possession (1 Cor. 10:19–21 and 1 Tim. 4:1). It is significant that while the apostle Paul lists healings (*charismata iamatōn*) in one of the lists of gifts that God has given to the church for ministry (1 Cor. 12:9, 28), he omits exorcism. Adolf Harnack found puzzling the absence of any reference to exorcism by the apostle Paul, given the frequency with which the subject appears in the Gospels.[36] But, in fact, apart from the Gospels, demons are seldom alluded to in first- or early second-century Christian literature.[37] The lack of mention of exorcism is perhaps best accounted for as reflecting the absence of demonic possession in New Testament churches. Christian writers of the first two centuries associated the widespread activity of demons in Jesus's time with his battle against the powers of Satan. They believed that after Jesus's resurrection demons continued to be present largely in the spiritual realm (e.g., as the active spiritual presences behind pagan idols).[38] Hence, although we find several incidental references to sickness in the canonical Epistles, they do not record a single instance of sickness that is either attributed to demons or healed miraculously.[39]

Let me summarize at this point what I believe to be the early Christian understanding of disease etiology as revealed by our necessarily rapid survey of the New Testament evidence. Apart from three apparent cases in the Gospels, the New Testament does not ascribe any illness or physical dysfunction to demons. Otherwise the description of illness is invariably that of common diseases or congenital physical impairments, none of them remarkable. For most of those diseases to which reference is made, no immediate causality is given, and their symptoms are clearly distinguished from those of demonic possession.[40] They have in common the absence of any symptoms that could reasonably suggest an immediate causality other than a natural one. Some illnesses in the New Testament remain unhealed and others are healed miraculously, while still others are healed, but whether by natural or miraculous means we are not told. There exists in the New Testament no condemnation of physicians or medicine, either specific or implied.

Demonic and Naturalistic Etiologies

In the archaic and classical periods (c. 750–323 B.C.) of Greek history, pagan religious thought regarded demons (*daimones*) as divinities who were subordinate to the gods. Plato's student Xenocrates developed a systematized demonology that distinguished between good and evil demons, and it became common in the Hellenistic age to attribute evil to demons rather than to the gods. Late pagan philosophers, such as Porphyry and Iamblichus, distinguished between the demons of popular religion, who received sacrifices, and the immaterial gods, whose worship was rational and superior. This distinction allowed demons to be increasingly identified in pagan belief with evil. Christian conceptions of demons might have been influenced by the writings of Philo, who transmitted Greek ideas to Judaism. Christians came to regard demons as evil spirits who were hostile to humans, and Christian apologists argued that the pagan gods were really demons, a view that became common among Christians. They regarded even healing gods like Asclepius as demons and attributed their healings to supernatural but evil forces.[41]

Ludwig Edelstein has demonsrated that demons had virtually no connection with Greek medicine,[42] which enjoyed widespread acceptance in Palestine in the first century.[43] It would be difficult, he observes, to find a physician who accepted a theory of demonic etiology. Physicians unanimously rejected it and treated the explanation with disdain.[44] Nor does Edelstein find any philosophers who accepted demonic etiology of disease, not even among the Neoplatonists of late antiquity who adopted thaumaturgy. Thus Plotinus (A.D. 205–269/70) heaps scorn on those who attribute disease to demons, arguing that disease is caused by natural factors and is healed by medical treatment.[45] One might expect physicians and philosophers, who were educated, to take for granted natural causation, but assume that popular opinion would incline to demonic explanations. But Edelstein argues that patients who believed in the demonic etiology of disease would not have sought the services of physicians because the latter attributed disease to natural causes. A doctor could not heal a disease that his patient thought demonically induced; the patient would have to consult an alternative healer who claimed to be able to expel demons by magic or drugs that they believed would be efficacious against them. If the sick consulted physicians, it was because they expected them to be able to heal diseases that they had diagnosed using natural explanations that their patients accepted just as they accepted their prescribed therapies.[46] "Tell me," writes John Chrysostom (c. 347–407), "if a physician should come to one, and, neglecting the remedies belonging to his art, should use incantation, should we call that man a physician? By no means."[47]

The evidence for pagan belief in demonic etiology in the first and second centuries is not extensive, but Plutarch (c. 50–c. 120) is sometimes cited as a contemporary source. In his essay *On Superstition* Plutarch contrasts the reactions to adversity of an atheist and a superstitious man. For Plutarch atheism was an insensitivity to the divine, while superstition was an emotional slavery to a distorted concept of the divine. Both are caricatures. Plutarch depicts an atheist, when ill, as attempting to recall a natural cause—an excess in eating or drinking or recent irregularities or vicissitudes—and seeking medical help. A superstitious man, by contrast, when beset by illness, places the responsibility for his lot on evil spirits or the gods. He sees himself as hateful in their sight, imagines that he is being punished by them, and acknowledges that his suffering is deserved because of his own conduct. When ill he will not consult a physician but will seek remedies from religious rites and purifications as he confesses his sins and mistakes.

Plutarch is not the only writer who dismisses as superstitious those who attribute illness to supernatural forces (e.g., gods, magic, or demons) and who seek remedies from purifications or incantations. The "superstitious man" was a familiar type in classical literature, one portrayed by Theophrastus (372/369–288/285 B.C.) in his *Characters* and by Lucian (born c. A.D. 120) in his *Lover of Lies*. E. R. Dodds thinks that Plutarch's depiction reveals a "religious neurosis" that is characteristic of the age, but John Gager and others have suggested that this view is overdrawn.[48] A specific application of superstition to medicine is made by Soranus (fl. A.D. 98–138), who, in describing the ideal midwife, writes: "She will be free from superstition so as not to overlook salutary measures on account of a dream or omen or some customary rite or vulgar superstition."[49] He does not mention attributing illness to demons, and perhaps one may infer that midwives did not need to be told that it was superstitious to do so. Aelius Aristides (A.D. 117 or 129–189), in spite of his obsessive commitment to following the cures recommended by Asclepius, diagnosed his ailments in the conventional terms of humoral medicine rather than of demonic etiology.[50] In fact, there is little evidence of educated pagans attributing any disease to demonic etiology,[51] and I have argued that there is almost no evidence that Christians did either. Dale Martin admits this lack of evidence but attributes it to the probability that the views expressed in our sources were class-based and that they reflect mainly those of the upper classes, but this is not a convincing argument for the view that only the educated classes accepted natural etiologies of disease.[52] While demonic etiologies are found in the magical papyri in Egypt, they date from the third through the fifth centuries and cannot be used as evidence for earlier centuries.[53] Finally, one cannot assume that a belief in demonic etiology of disease necessarily followed from a belief in demons. E. R. Dodds writes that Origen, "like nearly all Christians, believed in the reality and power of the pagan gods," which he (and they) considered demons.[54] We should not take his belief in demons as evidence, however, that either Origen or the Christian community in general held a demonic view of disease causation. In a culture where access to physicians who accepted naturalistic therapies based on some form of humoral theory was widespread, we can assume that most (though not all) people, both educated and uneducated, accepted at least a qualified natural explanation of disease.[55]

Early Christian Literature

When we turn from the New Testament to the earliest noncanonical writings of the late first and early second centuries (the Sub-Apostolic period), we find no more evidence of perceived demonic influence than we do in the New Testament Epistles. The writings that are conventionally referred to as the Apostolic Fathers were composed between c. A.D. 95 and 156 and follow without interruption the writings that came to be included in the New Testament canon.[56] They are chiefly pastoral and practical in nature, and we might expect them to address, or at least allude to, the activity of demons if demons were then regarded as afflicting Christians by means of possession or disease. Yet the writings of the Apostolic Fathers cite no case of either demonic possession or exorcism, much less of demonically induced illnesses.[57] Of course, our evidence is small, but it is unanimous. It is not until the apologetic literature of the latter half of the second century that we find a new and different emphasis on demonology that is starkly different from that which we see in the Gospels. The earliest Christian apologist was Justin Martyr, who was a convert to Christianity from paganism. Justin Martyr developed a system of demonology that he may have borrowed in part from pagan ideas. He is the first Christian author of the second century to mention healing and exorcism together as contemporary phenomena in the church,[58] although (as we have seen) the pairing of the two occurs in the Gospels and Acts. He writes that "many of our Christian men . . . have healed, and do heal" through exorcism, when other (non-Christian) exorcists failed to do so through incantations and drugs. It appears, however, that he refers to the healing of demonic possession (i.e., exorcism) rather than to the curing of illness that was caused by demons.[59] Whether Justin Martyr is using rhetorical exaggeration when he speaks of the existence of many Christian exorcists we cannot say. But other early apologists speak in the same fashion. Tertullian writes, "But we not only reject those wicked spirits [demons]: we overcome them; we daily hold them up to contempt; we exorcize them from their victims, as multitudes can testify."[60] I suspect that exorcism played a more important role in Christian rhetoric than it did in contemporary practice.

The attribution of disease and physical impairment to demons by the apologists is unclear and ambiguous.[61] Much of the difficulty in understanding precisely which view they held is due to their inconsistency. The gulf that separates the mentality of the twenty-first century from that of the early centuries of our era imposes yet another barrier to our understanding. Sometimes the apologists attribute disease and physical impairment to demons in a generic manner. Seldom, if ever, do they attribute specific cases to demonic causality. Thus Tertullian and Origen blame demons for disease and pestilence, as they blame them for all physical evils in the world, but only in a general sense.[62] At other times they maintain that demons do not cause disease or physical impairment but only simulate or feign it in order to deceive people into seeking healing from them.[63] Thus Minucius Felix (fl. c. A.D. 150) writes that demons who are worshiped as pagan gods "feign diseases, alarm the minds, wrench about the limbs; that they may constrain men to worship them being gorged with the fumes of altars or the sacrifices of cattle, that by remitting what they had bound they may seem to have caused it."[64] This explanation is commonly found in the writings of the apologists. Indeed, it constitutes a leitmotif of the demonology of the late second and third centuries.

Probably no Christian writer comes as close to attributing illness to demons as does Tatian, a second-century pupil of Justin Martyr's from Syria.[65] Tatian's theology was marked by a strong dualism and an ascetic flavor, and he is said by Irenaeus to have founded the sect of Encratites. Kudlien and others have asserted that Tatian rejected the use of medicine for treatment of illness, but as Owsei Temkin and Darrel Amundsen have demonstrated, he rejected only the use of drugs, which he believed to be a means of demonic intrusion. In fact, Tatian accepted natural herbal remedies but not compound drugs.[66] Although Tatian places a stronger emphasis than any other early Christian writer on the dangerous potential of demons to influence Christians, he sometimes says that demons merely feign disease while at other times he implies that they cause it.[67] His inconsistent language suggests that he was not sure (or perhaps we are merely unable fully to penetrate his understanding), but he seems to fall short of asserting a demonic etiology of disease. He does, however, follow Minucius Felix in believing that demons claim to cause disease in order to gain credit for its cure through the pagan healing cults. But, he warns, "the demons do not cure, but by their art make men their captives."[68] Tertullian expresses a similar sentiment.[69]

The apologists continue to speak of healing and exorcism in conjunction, but not as a single act, thereby reproducing the formula found initially in the Gospels. Thus Irenaeus (c. A.D. 130–c. 200) speaks of the gift of healing together with exorcism, prophecy, and the raising of the dead.[70] B. B. Warfield argued that, although Irenaeus appears to be describing contemporary phenomena here, he more likely has in mind miracles that were credited to the first-century apostles.[71] Several of the apologists speak of miracles of healing as events of the past that are no longer evident in their own day, although Origen writes that "traces of the [activity of the] Holy Spirit" are still preserved, which expel demons and effect many cures.[72] In fact, although the apologists describe Christian exorcism as a common event, they cite few examples from their own day.[73] Their failure to do so perhaps indicates that few were to be found.[74] Elizabeth Leeper suggests that the statements of Justin's regarding those who now believe in God, whereas they had formerly served demons, are made within the context of the power that Christ exercises over the demons that control the nations rather than of the power of exorcism.[75] It may well be that the second-century apologists prepared the soil in which later manifestations of possession and exorcism could flourish. This development seems not, however, appreciably to have influenced Christians' perception of disease etiology. Their continued acceptance of natural causation may be inferred from the high regard in which medicine was held by the apologists, which is not likely to have been solely a reflection of their educational background.[76] The evidence from the first two centuries, thin and scattered though it is, does not suggest that there existed among Christians an ideological or theological opposition to the use of medicine to treat disease.

Demons and Demonism in Late Antiquity

The phenomenon termed the "daimonization of religion" has been largely taken for granted in discussions of late antique religion, both pagan and Christian, as reflecting the climate of opinion in which Christianity grew. According to the conventional theory, in the third century "philosophy and folk-belief, instead of progressing along separate pathways, 'joined hands' in a common belief in demons which Christianity then transformed altogether into evil spirits."[77] This theory remains the currently dominant narrative, and it informs most descriptions of the intellectual atmosphere of the late Roman Empire.[78] Thus Peter Brown writes that a perceived ubiquity of demons was one of the elements of the "new mood" that spread rapidly in the age of the Antonine emperors (A.D. 138–180) and brought about the spiritual revolution that came to characterize late antiquity.[79] It has been taken to explain the fact that leading fathers like Origen and Cyprian (c. A.D. 200–258) make more frequent mention of healing and exorcism in the early third century than do Christian writers in the second.[80] The theory has, however, been challenged by several scholars as a narrative that is lacking in evidence. And once closely examined, it no longer seems very convincing.[81] Arthur Darby Nock argued that

Graeco-Roman society in late antiquity was not demon-ridden, or at least not more so than any other period of antiquity. Possession and exorcism were not new, while remedies, such as amulets and purifications, had long been readily available to guard against demons; in any event the perceived existence of hostile demons was no worse than that of the traditional gods, who had always needed to be appeased.[82] John Gager points out that Christians enjoyed no monopoly on exorcism and that Jewish or pagan exorcists were as available as were Christian ones.[83]

The question of when—if at all—the alleged "daimonization of religion" occurred is susceptible of several possible answers. (1) The second and third centuries witnessed a growth of belief in the activity of demons. Its growth was either (a) rapid (so Peter Brown, Valerie Flint) or (b) gradual, reaching its peak in the fourth and fifth centuries. (2) The second and third centuries saw no growth in belief in the activity of demons because belief in demons had long existed in the Graeco-Roman world (so Arthur Darby Nock, John Gager). Of these possibilities the theory that there was a gradual growth of belief in demons in late antiquity (1b) seems best to fit the evidence we have. The commonly held view of a sharp increase in the belief in demons in the second and third centuries relies less on contemporary evidence than on an assumed trajectory that traces the prominence of demons in the fourth and fifth centuries back to the earliest discussions among Christian apologists in the late second century. Peter Brown's assertion that the decisive factor in Christianity's growth was its superior ability to demonstrate "the bankruptcy of men's invisible enemies, the demons, through exorcism and miracles of healing"[84] lacks sufficient evidence to sustain it, particularly for the third century. Peter Brown and Ramsey MacMullen exaggerate both the ubiquity and the influence of demonic belief in late antiquity. Little evidence exists from contemporary sources to support the theory that Christians attributed disease to demons; in fact, that view assumes both that Christians adopted a Jewish demonic theory of disease etiology and that demonology grew rapidly in the second and third centuries.[85] Neither is supported by the evidence. Third-century medical theories of the etiology of disease were not demonic etiologies, and we have no reason to think that educated popular opinion differed from medical opinion in that regard.[86]

The church fathers' concern with demonic forces and the increasing use of exorcism in third century can be attributed in large measure to the popularity of the "Christus victor" (Christ the Victor) theme that emerged in the writings of the second-century apologists; certainly they cannot be understood apart from it.[87] Early Christian writers saw in Christ's redemption of the world his victory over the cosmic forces of evil that waged incessant warfare against God while at the same time afflicting humans.[88] This theme is found already in the New Testament, in which the victory over Satan appears as a prominent motif in the Gospel writers' portrayal of the ministry of Jesus. It is particularly demonstrated by his exorcisms, which they depict as mighty acts that foreshadow Christ's ultimate eschatological victory over the demonic kingdom. Thus in Luke 11:14–23 Jesus enters into a controversy with his critics over his casting out of a mute demon. When they accuse him of performing exorcisms by the power of Beelzebub, he replies that he expels demons by the finger of God and that by doing so he reveals the presence of the kingdom of God (v. 20). His analogy of the stronger man who overcomes (*nikēsēi*) the armed strong man and disarms and despoils him (vv. 21–22; the Synoptic parallels [Mt. 12:39 and Mk. 3:27] substitute "binds" [*dēsēi*]) suggests that he demonstrates his superiority to demons by his power over them. Similarly, Jesus attributes to Satan's fall from power the authority that his seventy disciples claim to have over demons (Lk. 10:17–20). Jesus is understood by patristic commentators to challenge Satan's power in another way by overcoming temptation (Mt. 4:1–11), not merely for himself but for all humanity in its struggle against demonic forces. But it is in his death on the cross that Jesus is seen by the fathers to win the decisive victory over Satan and, in his resurrection from the dead, to conquer death itself. The apologists saw in these events a reversal of the curse of sin that had enslaved the world under Satan's rule and the beginning of the destruction of the demonic powers. As Christ had been victorious over the powers of evil, so his people had been freed from the power of demons and overcame the world by their obedient faith (1 Jn. 4:4, 5:4–5). Like him they could defeat sin, demonic forces that were part of the world system (*kosmos*) that was Satan's, and—ultimately—death.[89] Michael Green's characterization of the period as "an age which was hag-ridden with the fear of demonic forces dominating every aspect of life and death," while typical of the traditional view (as is his exaggeration), must be regarded as built on tenuous evidence.[90] While Christians viewed the world in cosmic spiritual terms as an ongoing struggle between the forces of Satan and those of God, the theme of "Christus victor" provided assurance that their own victory had already been accomplished in principle. The defeat of Satan on the cross foreshadowed the ultimate victory of the Lamb at the end of the aeon (Jn. 12:31; Rev. 17:14). Even disease and death, which were elements of the evil that Jesus had overcome (Mt. 8:16–17 in fulfillment of Isa. 53:4–5), would someday be abolished (Rev. 21:4 in fulfillment of Isa. 25:8).

In the first two centuries Christians regarded the power to perform exorcisms as being divinely bestowed on the apostles and a select few.[91] The formula was simple: uttering the name of Christ together with prayer or adjuration was thought to drive demons out of persons or places. By the third century it became the practice to pronounce a rite of exorcism before baptism for the purpose of separating catechumens from the moral influence of evil, an account of which is given in the detailed description of baptism in the *Apostolic Tradition,* attributed to Hippolytus (c. 170–236).[92] Christians believed that converts, who had previously served pagan gods, had to be exorcized before formally entering the church as believers. They did not view them as possessed but rather as having been under the spiritual control of demons, which they believed the pagan gods to be. In undergoing exorcism catechumens were symbolically freed from their rule and acknowledged their own victory over demonic forces. The ceremony, which was administered to large numbers of converts from paganism, served a very different function than did the expulsion of demons in the case of those who were thought to be possessed.[93] In that sense, the difference between the exorcism of the Gospels and that of the third century is considerable.[94] The exorcism that developed in the third century required no drama or spectacular effects and was not even regarded as miraculous. It was a liturgical rite that sometimes incorporated anointing or the imposition of hands, after which the catechumen voluntarily renounced Satan.[95] The increased frequency of exorcism demanded the creation of an order of exorcists in the mid-third century.[96] Exorcists were made a minor clerical order, positioned between singers and doorkeepers. There is no reason to assume that the creation of an order of exorcists indicates the widespread use of the rite for healing from disease or illness. In fact, precisely the opposite was true. Christians saw in exorcism, much more than in miracles of healing, proof of the truth of Christianity that appealed to pagans.[97] Proclamation of the name of Christ in rites of exorcism carried with it the message of the defeat of demons and victory over Satan's kingdom.

In the fourth century new influences entered Christianity. Prominent among them were the rise of asceticism, the increasing resort to pagan or Christian magic, the veneration of relics, and the growing importance of holy men.[98] Yet the increased resort to miraculous forms of healing that followed did not lead Christians to abandon medicine, although it may have caused greater numbers of the ill or physically impaired to seek miraculous healing in addition to, rather than in place of, medical cures, or when medical healing had failed or was unavailable. Even with recourse to miraculous forms of healing, it appears that Christians typically continued to seek healing from physicians. This is indicated by the fact that while we find only scattered references to Christian physicians before the fourth century (probably owing to the paucity of our sources and the fact that Christianity was a *religio illicta*), thereafter they appear with regularity. Evidence that a naturalistic understanding of disease remained the predominant model among fourth-century Christians is provided by the list of miracles that Augustine (354–430) records in book 22 of the *City of God*.[99] If I have read his accounts of the incidents correctly, he records nothing other than a natural causality for every disease or physical impairment to which he ascribes a miraculous cure.[100] Indeed, he remarks that many of those who were healed had previously sought the aid of physicians. Although one would hardly call Augustine's list a valid statistical sample, it can reasonably be regarded as representative of the Christian perception of disease in the Latin West in the early fifth century. It suggests that, in spite of the growing interest in demonology among Christians of late antiquity, natural causation remained overwhelmingly the accepted Christian model of physical impairment and disease. In communities in the eastern half of the Roman Empire, where ascetics enjoyed popular support, demonic etiology and the accompanying miraculous healing may have gained greater acceptance, although even there the rise of hospitals furnishes evidence that medical treatment continued to be regarded as the most appropriate means of healing.[101]

Natural Causation and Religious Faith

Two conditions merit special mention, namely, epilepsy and mental illness. In many ancient societies abnormal psychic or mental states that could not be explained otherwise, particularly those that manifested bizarre behavior, were attributed in popular thought to demonic possession. But early Christian writers by no means viewed most cases of epilepsy or mental aberration as possession. Since the Hippocratic treatise *Sacred Disease* (probably written between c. 420 and c. 350 B.C.) had first offered a nonsupernatural explanation of epilepsy, various physiological explanations were developed by medical writers that provided a naturalistic understanding of the condition.[102] This medical tradition was so strong that it continued well into the Middle Ages, and while demonic possession and epilepsy were sometimes popularly confused in late antiquity, a number of the church fathers viewed epilepsy in medical rather than in supernatural terms.[103]

Several fathers were familiar as well with the medical literature on melancholia and incorporated a naturalistic understanding of it into their pastoral attempts to distinguish between spiritual anxiety or despair and clinical melancholia.[104] Jerome, for example, recognized that the monastic life sometimes produced melancholic madness (*melancholia*) and recommended medical treatment rather than spiritual counsel.[105] Peregrine Horden has assembled much evidence both from the fathers and from early Byzantine writers (especially the author of the seventh-century Life of St. Theodore of Sykeon) to demonstrate that early Christian writers regarded possession as a broad concept that was not always treated by exorcism.[106] They did not, even in the Byzantine era, attribute all insanity to demons: Horden cites a variety of writers who distinguished between what we might (speaking anachronistically) term "pychoses" and "organic disorders," on the one hand, and demonically induced conditions, on the other.[107] But even conditions that were diagnosed as possession were sometimes treated medically.[108] Physicians and laypeople alike were able to distinguish between possession and insanity that could be attributed to nondemonic causes.[109] Like Byzantine saints and doctors, they envisioned a "spectrum of [etiological] possibilities."[110] Horden believes, moreover, that scholars like Peter Brown overestimate the influence of saints who healed the sick by exorcizing demons. "I suggest rather," he writes, "that the Byzantine world did *not* pullulate with ascetics, that there *were* villages complete without a holy man, a resident 'outsider' who would resolve conflicts and cast out devils. A visit to such a figure required a particular decision and effort, and often a considerable journey. For very many of the possessed it was, in consequence, only rarely feasible."[111] A naturalistic understanding of melancholia was shared by a surprisingly wide spectrum of both Eastern and Western fathers. Their attempts to account for this condition (often given in the form of pastoral advice) tend not to be monocausal. They often invoked the same naturalistic explanations of melancholia and insanity that were given by medical writers, while at the same time suggesting an additional supernatural influence. In some cases they juxtapose physiological and demonic factors—for example, in stating that demons take advantage of mental afflictions to engage in spiritual attacks. The diagnosis could also reflect folk-cultural or medical assumptions based on the same symptoms.[112] Perhaps this approach reveals their essentially pastoral concern for the spiritual effects of melancholia.[113]

The evidence is overwhelming that a natural causality that was the inheritance of the Greek theoretical explanation of disease enjoyed widespread acceptance throughout the Roman Empire among pagans, Jews, and Christians even in late antiquity. It would be anachronistic to use the term "natural causality" in anything approaching the modern understanding of that term. Harold Remus speaks of the classical understanding of nature as "canons of the ordinary."[114] Nature (Gk., *phusis*; L., *natura*) represented what could be known from *ordinary* human experience. Hence the Greeks defined human nature by long-recognized and frequently observed behavior. The *extraordinary* represented that which was contrary to nature (*para phusin*). In Greek medicine *hugieia* ("health") was the term used to describe the normal condition of the body, while *nosos* ("illness") denoted a departure from nature.[115] The doctor attempted by treatment to restore the body to its natural state by an art that took its model from nature. When treatment failed, the patient might seek divine healing, which went beyond ordinary means of healing, that is, beyond the art of the physician. The classical world understood patterns of frequency and regularity in nature and used them as a backdrop to exceptional phenomena, some of which were described as *adynata* ("the impossible").[116] These categories were not limited to the learned; ordinary people understood how nature worked, not in a theoretical sense but from the experience of personal observation.[117] Some phenomena transcended their everyday experience: they were either extraordinary or inexplicable. They did not necessarily involve preternatural intervention and were not always given explanation, but they would provoke awe and wonder and in that sense were regarded as marvelous. Canons of the ordinary varied from place to place and age to age, differing according to status and education.[118] Yet they often transcended cultural, national, and religious differences. It was a shared understanding of how nature operated that permitted Greek medicine to be adopted by Jewish communities (although not without some opposition) during the Hellenistic period as compatible with Jewish theology.[119] The evidence of early Christian literature indicates that it was accepted by Christians as well. They readily accommodated their belief in God's providential activity in the world to a naturalistic explanation of disease, which they took for granted. So long as Christians looked upon illness as the result of natural (even if providentially determined) causes, they sought treatment in medicine. They broke bones and contracted diseases like their non-Christian neighbors. For common ailments they ordinarily consulted physicians or employed folk remedies. They considered prayer for healing, whether sought by natural means or by God's direct intervention, appropriate. In chronic and untreatable cases Christian piety advised patient submission to God's will.

The Early Christian Understanding of Disease

Most people in ancient Palestine, or in the Roman Empire, for that matter, did not think in terms of medical models. When they became sick, they sought remedies to restore health. But in order to appreciate the conceptual framework within which Christians accepted a natural etiology of disease, it is necessary to understand its theological underpinnings. The New Testament understanding of disease was not simple and schematic but a complex one that combined medical and theological components. Early Christians perceived disease within the theistic worldview that they inherited from Judaism. Most Christians believed that under God's ultimate and comprehensive sovereignty Satan had great, though highly circumscribed, power to work evil. Disease and impairment were one of the aspects of material (as distinct from moral) evil that resulted from the Fall of the human race into sin. Of course, most ordinary believers did not use theological language of that sort, but many of them viewed disease within a framework that was informed by those assumptions. While Christians thought disease to be generic in the human race, popular opinion often viewed it as God's retribution for personal or hereditary sin: it was the dominant theodicy of the ancient world. Yet on at least two occasions—in considering a man who was blind from birth and on being told of the Galileans whose blood Pilate had mingled with their sacrifices—Jesus explicitly refused to make a connection between personal sin and physical affliction as punishment.[120] The writers of the Gospels depict Jesus, like his contemporaries, as routinely using language that describes disease as a natural and ordinary phenomenon. One sees this in the striking account in the Fourth Gospel of his raising Lazarus from the dead (Jn. 11:1–44). In the narrative Lazarus falls sick, a fact that is repeatedly alluded to by the language of generic illness that is found in verses 1 (*asthenōn*), 2 (*ēsthenei*), 3 (*asthenei*), 4 *(hautē hē astheneia*), and 6 (*asthenei*). In verse 11 Jesus asserts that Lazarus is sleeping (*kekoimētai*) but that he will go to Bethany to wake him up (*exupnisō auton*). In verse 12 his disciples respond, "Lord, if he has fallen asleep, he will be all right."[121] They assume that sleep will help to bring recovery from a natural illness or that it indicates that he is already recovering. The language of the narrative consistently denotes a serious physical disorder that led to a natural death. One finds in the passage no assumption of demonic possession or of a supernaturally induced disease on the part of the writer or of those depicted in the episode. In verse 21 Lazarus's sister Martha assumes (as does her sister Mary in verse 32) that her brother died of the illness, which took its natural course and which only Jesus could have prevented by miraculous healing. Here, as elsewhere in the Gospel accounts, the most reasonable inference is that first-century Palestinians generally believed that ordinary disease was the result of natural processes. While the popular belief certainly existed that *some* disease might be attributed to demons, without specific mention of this possibility in the narrative, it is likely that a natural causality was assumed, even when no specified causality is given in the text.

Two New Testament passages, Luke 13:10–16 and Acts 10:38, have frequently been cited as evidence that early Christians considered demons to be directly responsible for illness.[122] But in neither of these passages is the cause of disease attributed to the immediate influence of demons. Rather, they ascribe disease and physical disability to Satan as the source of evil, suggesting that disease results from the material effects of sin on the human race.[123] The first passage, Luke 13:10–16, relates the case of a woman who was "with a spirit that had crippled her." After Jesus had healed her, he described her as one whom Satan had bound.[124] The second passage is found in Acts 10:38, where Peter says that Jesus "went about doing good and healing all who were oppressed by the devil."[125] No distinction is made here between disease and oppression by the devil. All forms of evil are traced to God's having given the world over to Satan, even if in a limited sense, while the verbs *euergetōn kai iōmenos* ("doing good and healing") form a unity.[126] This passage presents redemption as a power struggle between God and Satan by the pairing of *dunamei* ("with power") and *pantas tous katadunasteuomenous hupo tou diabolou* ("all who were oppressed by the devil"). Luke's view in both passages is theologically driven and should not be narrowed to a question merely of the etiology of disease. The language is too inclusive for this, and the writer employed it to describe the general deterioration of nature after the Fall, when Satan overpowered humankind. Both the New Testament and early Christian literature recognize all disease and affliction as the work of Satan, although they typically do not identify him as the immediate cause.[127] Cortes and Gatti maintain that the writers of the New Testament ascribe all illness, like possession, to demons, who are Satan's messengers.[128] In one instance, however, that of Paul's thorn in the flesh, it is the *thorn* that is depicted as the "messenger" (*angelos*) of Satan (2 Cor. 12:7–10). But the language is ambiguous in both the two Lucan passages, and a direct causal relationship between disease and demons simply cannot be demonstrated.[129]

Early Christians viewed disease and physical impairment as part of the natural order of a fallen world that was under the dominion of sin and yet providentially ordered by a sovereign God. Christians valued medicine as God's gift for the natural healing of disease. But for a Christian to seek medical healing apart from dependence on God, the Ultimate Healer, was to substitute reliance on medicine alone for a Christian approach. Augustine had this in mind when he asserted that some Christians were too dependent on physicians because they clung too dearly to life.[130] Yahweh's words to Moses, "I am the Lord who heals you" (Ex. 15:26), remained for Christians, as they did for Jews, the foundation of their understanding of healing.[131] Many Christians could have quoted with approval the words of ben Sira: "Honor the physician with the honor due to him, before you need him. . . . [For] he [Yahweh] gave skill to men that he might be glorified in his marvelous works. By them he heals and takes away pain. . . . My son, when you are sick do not be negligent, but pray to the Lord and he will heal you."[132] Normal recourse to physicians (or traditional or home remedies) was the inevitable outcome of a naturalistic understanding of disease. But the church fathers repeatedly pointed out that it is God who heals through the physician. One's faith, cautions Ambrose (c. 337–97), should be in God, not in medicine.[133]

Conclusion

The attribution to early Christians of a demonological explanation of illness reflects a misunderstanding of the early Christian perception of disease causation based on a sharp distinction that some scholars have drawn between early Christian and naturalistic concepts of disease etiology, whether ancient or modern. Thus Robin Lane Fox, while not attributing a consistent demonic etiology of disease to early Christians, argues that they rejected naturalistic humoral etiologies of health. They rather "connected health with faith and an absence of sin. . . . Their image of the human body conformed to their image of the Church as a sinless Body, set apart from the demonic world."[134]

Our study suggests, rather, that early Christians, like the majority of their contemporaries, implicitly accepted a natural causality of disease within the framework of a Christian worldview. If they sometimes spoke in a manner that blurred the distinction between ultimate and proximate causation, it was because they believed that the presence of God was operative in natural forces. They viewed Jesus's exorcisms and miraculous healings as signs that the kingdom of God had come, not as normative models for the healing of ordinary disease. They sought out physicians for their diseases and valued the healing power of medicine. In their view, however, medical treatment and prayer were not mutually exclusive but necessarily complementary. Of course, some early Christians resorted to the use of amulets or relied on dreams, predictions, and portents, not because their faith encouraged them to do so (in fact, it explicitly forbade some of them) but because they were commonly appealed to in the larger culture of the Roman Empire. And where treatment had proven ineffective or few doctors were to be found, some would have had recourse to parallel therapies, consulting healers who employed magical or folk cures.

I am not arguing that a good deal of medical thinking that lay outside the etiologies and therapies offered by Graeco-Roman medicine was not to be found in early Christian communities or that the sick wholly subordinated their understanding of disease—of its causes and its explanation—to that of physicians.[135] But one finds a persistent reluctance among some modern scholars to admit that early Christians accepted disease as a natural phenomenon. In large part it stems from a failure to recognize that Christians shared the same climate of opinion as their pagan neighbors and employed the same medical categories as they did in diagnosing sickness and its causes. We are often told that Christians lived in a world that was filled with spirits and inhabited by baleful demons, with the implication that they sought supernatural explanations for their illnesses. In fact, where we have detailed evidence, it suggests that they talked about their illnesses, no doubt frequently, but in a matter-of-fact way, because they suffered more commonly than we do today from all kinds of diseases and physical disabilities.[136]

Anne Merideth cites the Cappadocians (Gegory of Nyssa, Basil, and Gregory Nazianzus) as examples of this focus on their illnesses because they wrote a great deal about them to each other (as did Marcus Aurelius and Fronto in an earlier age).[137] They described at some length their symptoms and discomfort in the illnesses from which they suffered, but they rarely spoke of their attempts to find a larger meaning in them. Moreover, they take for granted the typical medical explanations of disease, mostly in terms of humoral theory and the balance or imbalance of bodily fluids. In this they do not differ from non-Christians in their explanations of disease. The reason is that there existed no gulf in medical opinion between educated laypeople and physicians or medical writers.[138] Indeed, there was no specifically Christian approach to diagnosis, just as there was no specifically Christian means of healing, such as prayer, or exorcism, or unction. This is not to say that each of these was not sometimes appealed to by Christians. There has probably always existed a minority within Christianity that have claimed that God heals directly and without means. Moreover, it is important to distinguish in early Christian writers between the etiology that they assign to a particular disease and the ultimate meaning that they give to it, the latter being the result of meditation on its purpose. Reflective Christians would often have asked themselves why they were suffering or why God was afflicting them, as believers—and nonbelievers—in every society have asked. But one can seek ultimate meaning regarding human experience while at the same time taking it for granted that ordinary diseases have natural causes, which can be accounted for by whatever medical models are dominant in any culture.

Early Christians did not on the whole contrast moral with physical causes of sickness, though some did, particularly within the ascetic tradition in late antiquity.[139] Scholars who assume either the normative character of modern medicine or the explanatory models of medical anthropology emphasize the gulf between early Christian and naturalistic understandings of the cause of disease and its cure. I suggest that, with a greater appreciation of the social context of illness and its definition, our own perception of disease is as much conditioned by contemporary medical models as the early Christians' perception was by the ordinary medical models of their age and that both reflect the climate of their times. This is true irrespective of their belief, or ours, in the relation between a supernatural world and the everyday affairs of men.

Christianity as a Religion of Healing

Since the time of Adolf Harnack (1851–1930) it has been widely maintained that an emphasis on physical healing was, from the New Testament era to the end of antiquity, a major aspect of early Christianity.[1] One might cite many authorities for this view.[2] I merely adduce two. First, Harnack: "Deliberately and consciously [Christianity] assumed the form of 'the religion of salvation or healing,' or of 'the medicine of soul and body,' and at the same time it recognized that one of its cardinal duties was to care assiduously for the sick in body."[3] In a somewhat different vein Shirley Jackson Case writes: "In the ancient world it was almost universally believed that the function of religion was to heal disease, and it was in just this world that Christianity took its rise. It need not surprise us, therefore, to find that Christianity is from the start a healing religion."[4] That this view has gained something of the status of an orthodoxy is evidenced by a statement made by Vivian Nutton, who referred to Christianity as "a healing religion *par excellence*" and suggested that "this was one of the features that secured for Christianity the primacy among competing religions."[5]

Precisely what kind of healing did early Christians claim to offer? Healing of the body, or of the soul, or of both? And if, as has often been maintained, their offer of healing included the healing of the body, what sort of healing did they have in mind? Physical healing through ordinary means, or miraculous healing? Most scholars who have stressed the role of healing in early Christianity have emphasized the latter. Again I quote Vivian Nutton: "Yet this Christian healing was not that of the doctors. It succeeded where they had failed, often over many years and at great expense; it was accessible to all; it was simple. It was a medicine of prayer and fasting, or of anointing and the laying on of hands."[6]

It is this thesis—that early Christians offered an alternative model to secular healing, the substitution, or at any rate expectation, of religious healing for the use of medicine[7]—that I should like to examine in this chapter, first by considering the evidence of the New Testament, then by looking at Christian writers of the second century in some detail, and finally by contrasting the first three centuries of Christianity with the late fourth and fifth centuries regarding Christians' attitudes toward healing. I shall argue that religious healing enjoyed little prominence in the first three centuries and that there is evidence of a major shift in emphasis—in which religious healing secured a prominence that it had not attained earlier—during the late fourth century.

The New Testament

There is much prima facie evidence for the view that miraculous healing played an important role in early Christianity. If we turn to the New Testament, we find many accounts of miraculous healing by Jesus.[8] David Aune cites seventeen instances of his healing the blind, the deaf, the dumb, the lame, lepers, and the disabled of all sorts.[9] There is no doubt that miracles of healing are assigned a central place in the ministry of Jesus by the writers of the Gospels, who consistently portray them as a manifestation of the presence of God's kingdom. The miracle narratives are not, however, presented as if they were an end in themselves. Rather, they represent the external aspect of salvation, the physical manifestation of a new spiritual order.[10] The vocabulary of the Gospels (both the Synoptics and the Fourth Gospel) is revealing.[11] The healings performed by Jesus are spoken of as "signs" (*sēmeia*) that bear witness to his messianic credentials, and they are regarded as the fulfillment of prophecies contained in the Hebrew scriptures (see, e.g., Mt. 11:4–5, which echoes Isa. 35:4–6 and 61:1).[12] Thus Matthew 8:16–17 describes Jesus's ministry of healing and exorcism as a fulfillment of the prophecy of Isaiah 53:4 ("He took our infirmities and bore our diseases"). Jesus himself is said in the Fourth Gospel to have cited his miracles as a sign of his messiahship (Jn. 10:37–38; cf. Acts 2:22).[13]

In the book of Acts we find several accounts of individual miracles of healing that have much the same character as those performed by Jesus. These acts of healing are always attributed to the apostles, who had been for the most part the disciples of Jesus, but to no one else. The reason for this is that the apostles, like Jesus, were believed by the Christian community to have their ministry accompanied by signs that served to confirm their apostolic credentials as authoritative agents of God (see 2 Cor. 12:12).[14] Thus we read of Paul and Barnabas that they "remained for a long time, speaking boldly for the Lord, who testified to the word of his grace by granting signs and wonders to be done through them" (Acts 14:3). Only a relatively small number of healing miracles are attributed to the apostles in Acts.[15] Moreover, it is interesting to notice that in their preaching they make little reference specifically to the healing power of Jesus (although in one instance [Acts 10:38] Peter does refer to it). Thus in Acts 3:8–10 Peter and John are said to have healed a man in Lystra who was lame from birth. Shirley Jackson Case argues from this and similar passages that "the early Christians were not at all inferior to their coreligionists [*sic*] in professing to possess power to heal all kinds of diseases."[16] Yet in a speech to the assembled crowd after a healing at Lystra, Peter speaks of the healing as a sign that salvation has come. He does not call upon those assembled to be healed of their physical ailments but rather to repent.[17] The theme of the apostles' preaching is salvation through Jesus Christ, and miracles performed in his name exhibit his superior power. This is equally true whether they were preaching to pagan or to Jewish audiences, a notable fact given the obvious appeal that a new religion of physical healing would have had in the pagan world of healing cults.[18]

If we turn from the Gospels and Acts to the Epistles of the New Testament, we find little specific mention of healing.[19] The apostle Paul does indeed mention the subject in two lists of the gifts that the Holy Spirit has given to the church (1 Cor. 12:9 and 28). But he does not, here or elsewhere, elaborate on it or describe how it manifested itself. The Epistles are our earliest records of Christianity, having for the most part been written before the Gospels. They were meant by their authors to provide normative apostolic teaching on matters of faith and practice for the churches to whom they are addressed. If in its earliest phase Christianity emphasized healing, we should surely expect to find evidence of it in the Epistles. They contain, however, only one discussion of the practice of healing, which is found in the Epistle of James, perhaps the earliest of all the New Testament writings, composed quite possibly before A.D. 50.[20]

The Epistle of James prescribes a rite of healing in which the presbyters (or elders) of the local congregation anoint the sick and pray for their recovery, which is assured (5:14–16). The passage is a difficult one, which has been interpreted in a variety of ways.[21] Does it refer to physical health and recovery or to a spiritual condition? Although the weight of scholarly opinion is in favor of physical healing, one might raise some compelling arguments in support of the view that spiritual healing and restoration are the object. A few verses earlier James cites Job, who patiently endured his physical affliction, as an example for the Christian's emulation

(5:10–11). There are, moreover, intimations in Acts and the Epistles that not all Christians were miraculously healed—or healed at all—in the apostolic age: Paul's friend Epaphroditus (Phil. 2:25–27) and Trophimus (2 Tim. 4:20), for example.[22] We cannot say whether either of these cases resulted in healing by miraculous means or even (in the case of Trophimus) whether he recovered. Even Paul continued to suffer from his "thorn in the flesh" (2 Cor. 12:7–10), and though it was probably a physical ailment, we cannot be sure what it was.[23] The vocabulary employed in James suggests, rather, that the promise of healing might be addressed to those who are spiritually, not physically, ill. While the verb *kamnein* (v. 15) can mean "to be ill," it is more often used for fatigue or weariness in New Testament literature (e.g., in Heb. 12:3). The meaning might be that those who have lost the joy of their salvation (see the contrast to v. 13b) should call on the help of the community (i.e., elders). The spiritual fatigue is left unqualified, but the addition of verse 15b makes explicit that if a particular sin is its cause, it will be forgiven. The juxtaposition of *sōsei* (v. 15) and *iathēte* (v. 16) might argue for a figurative interpretation of the passage in light of the frequently found pairing of salvation with health in biblical literature.[24] The order in which they appear seems to be the reverse of what one would expect: the prayer of faith will *save* (*sōsei*) the sick, while those who confess their sins will be *healed* (*iathēte*). The sick appear to be the spiritually ill, while healing (in this case, perhaps forgiveness and reconciliation with God) is given not for a physical condition but to the sin-sick soul. The text does not demand that the presbyters have access to supernatural power in performing the rite; rather, it emphasizes the power of prayer exercised in the regular ministry of the church. Hence I suggest that this passage should not be taken as providing the basis of a rite of miraculous healing in the early church that was routinely administered by the presbyters, especially given the fact that it stands alone in the New Testament Epistles as a possible warrant for the expectation of miraculous healing.

Unction for healing is mentioned in the Gospels as having occurred during Jesus's ministry (Mk. 6:13), but we do not, outside James, read of its use in the first-century church. Anointing in the New Testament Epistles is often used figuratively.[25] The first recorded mention of the use of unction for healing dates from the early third century, when Proculus Torpacion, a member of the Montanist sect,[26] is said to have healed the Roman emperor Septimius Severus (193–211) by anointing.[27] The passage in James is referred to by Origen[28] and John Chrysostom,[29] but only the latter connects it with physical healing.[30] It appears that it was not until the fourth or fifth century that anointing of the sick became a part of the sacramental liturgy that was administered by presbyters and bishops in churches.[31] Yet if the passage was seen by early Christian readers to be describing *spiritual* healing, as Origen read it, that fact might help to account for the lack of perceived warrant in the early church for the practice of anointing for *physical* healing. In any event, the silence of our early sources makes it doubtful that the practice was widely used before late antiquity. The connection between sin and sickness was widespread in Near Eastern cultures, and as ritual healing developed in Christian communities, some practiced anointing for the forgiveness of sins, and others offered it for the cure of the body, while some did it for both.[32]

The Second Century

If some sort of sacramental or ritual healing was employed continuously in the early church, or if miraculous healing of any sort enjoyed prominence, we should expect to find mention of it in the second century. B. B. Warfield long ago observed that, in contrast with Christian writings of the fourth century, those of the second are largely lacking in references to contemporary miracles of healing.[33] This is certainly true of the earliest noncanonical writings of the late first and early second centuries, which are conventionally referred to as the Apostolic Fathers (e.g., the Epistles of Clement, Polycarp, and Ignatius). These writings are chiefly concerned with the internal life of the Christian communities. While the sample is small, it may be significant that we find in them no specific mention of contemporary healing practices.[34] In contrast, when we turn to the apologetic literature of the latter half of the second century, we find reference to healing in the writings of Justin Martyr, Irenaeus, Origen, and Tertullian.[35] Justin Martyr, for example, mentions healing and exorcism together as a contemporary phenomenon in the church.[36] Irenaeus (c. 130–c. 200), too, speaks of the gift of healing by the imposition of hands that has been given to Jesus's "true disciples"; this gift, he says, the church employs "day by day for the benefit of the Gentiles," together with exorcism, prophecy, and raising the dead.[37] This statement by Irenaeus is the most explicit reference to the continuing claims for miraculous healing that we have from the second century. What is remarkable about these references, however, is that they are invariably couched in general terms: their tone is that of conventional apologetics, and they cite no specific instances of healing.[38] Hence, although they appear to speak of contemporaneous phenomena, they more than likely refer to miracles that were ascribed to the apostles in the New Testament.[39] It is noteworthy that Irenaeus's allusion to miraculous healing is found in the context of a refutation of heretics, where he is concerned to distinguish between true and false miracles.[40] It may well represent a defensive rhetorical strategy on the part of the apologists.

Although Irenaeus implies that miraculous healing was a contemporary phenomenon, several writers of the second and third centuries suggest that the abundance of miracles attributed to the apostolic age was no longer to be seen in their own day. Thus Origen, after speaking of the miracles performed by Jesus and the apostles, writes that there are still preserved among Christians only "traces of the Holy Spirit" that expel demons and effect many cures.[41] Similarly, Tertullian cites the raising of the dead and the healing of disease as examples of miracles performed by the apostles, which are not (he implies) shared by their successors.[42] Irenaeus, in speaking of those who were raised from the dead, places the events in the past tense, suggesting that he too regarded his age as somewhat different from the apostolic era insofar as the existence of supernatural gifts was concerned.[43]

An additional argument for the lack of belief that miraculous healing still occurred in the church in the second and third centuries is the failure of Christian writers to exploit for apologetic purposes specific cases of Christians who claimed to have been miraculously healed. The desire for physical health and healing was widespread in the second century. Diseases of the body and the mind, real and imagined, abounded, and with them the demand for miraculous cures, a demand that centered on the many healing cults throughout the Mediterranean world.[44] There were hundreds of pagan healing shrines in the second century, some connected with local hero cults, others with Asclepius, who had emerged as the healing god par excellence of Graeco-Roman culture and who enjoyed extraordinary popularity in the age of the Antonines (138–80).[45] Other gods with a claim to universal worship (e.g., Isis and Serapis) attracted large numbers of people seeking relief from disease.[46] Testimonies to miraculous cures effected by the gods (called aretalogies) were employed to attract those seeking healing. The eagerness for healing produced as well a crop of charismatic healers such as the "pagan Christ," Apollonius of Tyana.[47]

Ramsay MacMullen has repeatedly argued that of all the elements of Christianity that appealed to pagans, miracles of healing by exorcism had the greatest attraction.[48] "It could thus be only a most exceptional force that would actually displace alternatives and compel allegiance; it could be only the most probative demonstrations that would work. We should therefore assign as much weight to this, the chief instrument of conversion, as the best, earliest reporters do."[49] MacMullen believes that Christians "stood out as frequent and powerful exorcists," even though no great audience or reputation for exorcism existed among pagans.[50] In fact, the "earliest reporters" do not emphasize exorcism or miracles of healing (for MacMullen they are usually the same) in the way that he thinks they do. Elizabeth Leeper observes that it is in the Christian romances and legends that we see "all the detailed stories of exorcism and conversion . . . the apologists give no details, no examples, not a single named individual who saw an exorcism or was him- or herself exorcized and became a Christian thereby."[51] Until the time of Ambrose and Augustine in the early fifth century, healing miracles were seldom exploited in apologetic literature. The early apologists were anxious to counter the claims of Asclepius, not by disproving them but by demonstrating that his healings were the result of demonic forces.[52] While they assert that Christians too can demonstrate cases of healing, they place little emphasis on them. To the contrary, their general defense of miracles does not suggest personal experience with actual miracles, and any claim that they were influential in winning converts has to take into account Origen's statement that they were becoming less common with the passage of time.[53] Without explicitly stating that the age of miracles had ceased with the apostolic era, the apologists hint broadly that they have only a few traces of miracles in their own day. Even Irenaeus, who asserts the continuing presence of miracles,[54] like his contemporaries, speaks generally and without citing examples. But Eusebius, who quotes him, suggests that in Irenaeus's time, as in his, they were not common.[55] Origen, rather, emphasizes the importance of visions in conversions,[56] while the only kind of miracle that remains prominent in apologetic works is exorcism.[57]

Robin Lane Fox argues that by crediting miracles or exorcisms with winning converts we "shorten a long process" and underestimate the "extreme canniness of Mediterranean men."[58] The Roman world was used to claims of miracles, and the educated classes thought exorcism "tommy-rot." It was necessary in the milieu of pagan cults not to win adherents, but by conviction and persuasion to win converts. Miracles played little role in gaining a hearing with pagans; reasoned argument did. "We know of no historical case," writes Lane Fox, "when a miracle or an exorcism turned an individual, let alone a crowd, to the Christian faith."[59] The point is nicely illustrated by a revealing discussion of miraculous healing from a Christian point of view that is found in Origen's *Contra Celsum,* written in the third century as a rejoinder to an attack on Christianity by the learned late second-century pagan Celsus. A disputed point concerns the question whether Asclepius or Jesus is the true savior. One might expect Origen to cite instances of divine healing in Jesus's name, particularly given the numerous claims of divine healing attributed by Celsus to Asclepius. In fact, he cites none. Indeed, Origen's argument is rather weak. He is prepared to admit the healing power of Asclepius. Early Christian writers like Origen believed that demons could heal as well as God could and that Asclepius was not a god but a demon who used magic to heal. But he says little in support of the healing power of Jesus, except to claim its existence.[60] One has the feeling that when claims of contemporary miraculous cures were put forward in the second century, in debate between the followers of Jesus and those of Asclepius, Christians discovered themselves to be at a disadvantage. They found it difficult to discredit Asclepius, whose cures were abundant and whose claims were hard to deny, let alone to match.[61] That the Christians appear to have made little attempt before the late fourth century to match them argues that healing (as distinct from the care of the sick) enjoyed a lack of emphasis in Christian circles. Perhaps even some Christians were not immune from the attraction of Asclepius. One finds in Christianity in the second and early third centuries—an age that eagerly sought religious healing—a notable lack of the elements of a healing religion: evidence of few healers to compete with the pagans; and little evidence of attempts to use specific instances of miraculous healing for apologetic purposes, or even to counteract the claims of the pagan cults. Finally, in an age in which Christianity was growing at a significant rate, we hear (pace MacMullen) of few proselytes brought into the faith either because they had been healed or because they expected to be. The popular appeal of miracles of healing was less important in securing Christian converts than were argument, persuasion, and a theology that brought conviction and hope to those who accepted it.[62]

Healing in Early Sectarian and Heretical Movements

One might expect to find more evidence of a tradition of miraculous healing among tangential or sectarian groups on the margins of Christianity, since there is so little evidence of it within the mainstream church. Montanism, which claimed for itself the continuing existence of the prophetic gifts exercised by the apostles, seems a promising candidate in this regard.[63] Montanism was named after its founder, Montanus, who hailed from Phrygia in western Asia Minor and who claimed to possess the gift of prophecy. In either 156–57 or 172 he began to utter direct revelations from God through the alleged inspiration of the Holy Spirit.[64] The gift of prophecy was also claimed by two of Montanus's female followers, Priscilla and Maximilla.

Unlike mainstream or Catholic Christians, Montanists claimed that the supernatural phenomena (*charismata*) manifested by the apostles continued into their own day as a part of the permanent ministry of the church. These *charismata* included not only ecstatic prophecy but also visions and glossolalia (speaking in tongues). Montanism spread from Asia Minor to North Africa, where it enjoyed widespread appeal, but we do not know how many of its adherents actually left mainstream churches to form distinctly Montanist ones. Montanist influences led, however, to the appearance of ecstatic prophecy as a familiar phenomenon in North African churches. Tertullian, who himself became a convert to Montanism later in life (a Montanist tendency begins to be apparent in his writings in 207), speaks of such a case in his *De anima*, which was written during his Montanist period, in the context of a passage in which he deals with the subject of the *charismata*. He writes of "a sister whose lot it has been to be favoured with various gifts of revelation, which she experiences in the Spirit by ecstatic vision in church during the sacred rites of the Lord's day: she converses with angels, and sometimes even with the Lord; she both sees and hears mysterious communications, she understands some men's hearts, and she distributes remedies to those who are in need."[65] The woman is not described as a Montanist, but the very fact that as a woman she exercised a prophetic ministry may suggest a Montanist influence.[66] The reference to her distributing remedies to those who ask may refer to religious healing, but the passage does not make this clear, and it is a very slender thread on which to hang the conclusion that religious healing was practiced by Montanists. Prophecy, visions, and glossolalia are specifically attributed to the Montanists by our sources, but healing is not. In fact, Montanus and his prophets did not claim to perform miracles.[67]

The Montanists apparently employed incubation for visions and revelations in Asia Minor (i.e., receiving visions as dreams),[68] and one should not exclude the possibility that they employed it for healing as well, especially given the fact that incubation was employed in the temple of Asclepius at nearby Pergamum. But there is no indication that incubation was ever practiced for healing at the village of Pepuza, which was the Montanist religious center in Asia Minor. Montanus, Priscilla, and Maximilla were buried together there, and their remains, like those of Elisha, came by the sixth century to have healing powers attributed to them.[69] But we have no evidence that the site attracted pilgrims who sought healing in the second century. Nor does the fact that a Montanist theologian, Proculus Torpacion, healed the emperor Septimius Severus by means of unction tell us much about the specific healing practices of the sect, if indeed there were any. If miraculous healing did play a role in Montanism, we should expect to hear more of it. But the evidence is insufficient to infer that it accompanied the charismatic phenomena that we know were claimed by the Montanists.

The evidence for miraculous healing in other early Christian sects that came to be branded as heterodox suggests that, while some claimed it, their success was slight. According to Irenaeus, the Gnostics, as well as "other so-called workers of miracles," deceived the faithful by magic.[70] But Irenaeus denies that they restored sight to the blind or hearing to the deaf or cast out demons, as they claimed to do. It was in the second century, however, that sects such as the Gnostics (religious and philosophical movements, some claiming to be Christian, that flourished from the second to the fifth centuries), the Ebionites (a Jewish Christian sect), and the

Encratites (an ascetic Christian sect) began to produce apocryphal gospels and acts that describe fanciful and bizarre miracles, including miracles of healing, which they ascribed to Jesus and the apostles.[71] While the number of new miracles that are ascribed in these gospels to Jesus as an adult are few (only three, on Paul Achtemeier's count), miracles ascribed to him are more frequent in the infancy gospels. But those ascribed to the apostles are abundant in the apocryphal acts.[72] In many cases (e.g., the infancy gospels, which purport to describe the childhood of Jesus) these works take the form of novelistic romances, based on pagan Graeco-Roman models, that appealed to popular curiosity regarding the "hidden years" of Jesus's life.[73] The tendentious character of many of them, however, reflects one or another sectarian parti pris. Perhaps proponents of some early heterodox view intended to furnish support for their own claims of miraculous healing by creating pseudepigraphical texts that magnified the role of such healing in the first century.[74] Yet stories of the disciples' raising people from the dead are more numerous than are those of healing.[75] Achtemeier suggests that the major theme of the Acts of Peter, a late second-century text that is one of five major extant apocryphal acts, is "the emphasis on the ability of the apostle to win contests of miraculous power." In the competition of truth claims, the greater the miracle, the more credible the claims of faith become. Thus in the contest between Peter and Simon before leading officials in Rome, the superiority of Peter's miracles demonstrates the veracity of his religious claims, as a result of which the crowd stones Simon and believes in Peter.[76]

Judith Perkins argues that a new concern with health that appeared in the second century represented a cultural transformation in the perception of the human body, which increasingly focused on physical suffering rather than on healthy bodies.[77] Perkins argues that this concern gave rise to early Christians' discourse on pain and suffering that was socially constructed and intended to create a need for its own ideology and for the services that it could offer.[78] Christians attempted to meet the demand for health by offering healing[79] and release from death. Perkins cites as evidence the Acts of Peter,[80] the author of which, she argues, intends among other things to "establish the superior healing prowess of the Christian community." But she confuses the metaphor of Christ as physician of the soul with that of the body.[81] The Acts of Peter, a tendentious romance containing Gnostic elements and Encratite tendencies, is simply insufficient evidence on which to base a claim for "the Christian community's powerful concern with sickness, health, and human suffering."[82] Its tone differs from that of much mainstream contemporary Christian literature, although it is of a piece with the themes developed in the Gnostic gospels and acts that proliferated in the second and early third centuries. Averil Cameron describes their "recognition scenes, travel narratives, wonders, young girls in trouble," which are closely related to novelistic literature.[83] But the Acts of Peter also differs *toto caelo* from its ostensible model, the canonical book of Acts. The latter is similar to the Hebrew scriptures in its approach to miracles: it views them as *semeia kai terata* ("signs and wonders") that reveal the hand of God in bringing about the salvation of Israel. The Acts of Peter, in contrast, depicts conjuring tricks whose world is that of Hellenistic wonder-workers. Thus when Peter carries on a contest in miracle working with Simon Magus in Rome, he throws a sardine into a pool and causes it to swim, not merely for a short time (which might indicate a trick on his part) but long enough to eat bread thrown to it by the crowd, thereby amazing large numbers of observers, who immediately convert to Christianity.[84]

One's understanding of Christian attitudes toward healing and medicine in the second and third centuries depends to a great extent on which sources are used.[85] Perkins and other scholars argue that the Acts of Peter and similar apocryphal works provide evidence for the widespread popular belief of second-century Christians in the ubiquity of miracles, demons, spirits, healings, and exorcisms.[86] Do we privilege the apologists, who demonstrate a respect for medicine and regard the Christian's use of it as a gift from God, or do we take the pseudepigraphical gospels and acts as indicative of the mentality of the Christians who read them? While scholars like Perkins take the second view, it is questionable whether the heightened supernaturalism and fascination for the sensational of the apocryphal literature characterized the mainstream Christian community. The Acts of Peter may have found numerous readers outside Gnostic circles, given the fact that its author sought to adapt his version of Christianity to contemporary literary tastes.[87] But its theological influence is likely to have been limited by the fact that Gnostic beliefs were routinely and strongly attacked by orthodox apologists, who warned the Christian community of their heterodox ideas. Like later hagiographical literature, these Christian romances exalted heroes of the faith, which provided encouragement in time of persecution. The characters were not contemporary Christian saints, however, but Jesus's family and disciples, around whom apocryphal tales had gathered. There is no evidence that they preserve any historical material that we do not already find in the Gospels. G. W. H. Lampe calls them "a sort of equivalent to science fiction, and they plainly belong to the sphere of fantasy."[88] Their readers probably believed that miracles of the sort that they described were possible, especially when they attributed them to Jesus's disciples. And they certainly believed that they demonstrated the superiority of the God worshiped by the Christians to the pagan gods.[89] But the world of marvel and fantasy depicted in the Acts of Peter is very different from the mentality exhibited by more mainstream Christian writing of the second century. One can believe that the literature that described it circulated in Christian communities, both Gnostic and Catholic, while doubting that it encouraged them to expect miracles and marvels of the sort that they found described in the apocryphal acts. These works are more likely to have met the need for sensational literature that every era experiences than to nurture the credulity of the Christians who read them. Like the apologists, they viewed the apostles' miracles as demonstrations of the claims of Christianity but found the appeal of the marvelous more inviting than the claims of reasoned apologetic, as they did hagiographical literature in subsequent centuries.[90]

It is sometimes assumed that those who read the apocryphal gospels and acts belonged to a different social class than those who read the works of the apologists. But to correlate intellectual ability with class is always problematic. Robin Lane Fox points out that theological debate was carried on in the second century at a high level by relatively humble Christians. "Christianity," he writes, "made the least-expected social groups articulate." They were attracted to theological discourse "not because of miracles but because the ideas appealed to them."[91] Even simple Christians could be attracted to schisms and heresies, not for their miracles but for their theological ideas. Gregory of Nyssa's (c. A.D. 330–c. 395) description of the widespread interest of his fourth-century Christian contemporaries in theological speculation is well known: "If you ask about your change, the shopkeeper talks theology to you, on the Begotten and the Unbegotten; if you inquire the price of a loaf, the reply is: 'The Father is greater and the Son is inferior'; and if you say, 'Is the bath ready?' the attendant affirms that the Son is of nothing."[92] According to Lane Fox, "Christianity's theology combined simple ideas which all could grasp but which were also capable of infinite refinement and complexity."[93] The pagan critic of Christianity Celsus described its leaders as "wool workers, cobblers, laundry-workers, and bucolic yokels."[94] While he was by no means a friendly critic, he may not have been far off the mark. Theodotus, who read Galen and founded the Monarchian heresy, was a leather worker. Yet interest in popular philosophy was widespread among pagans too. It reflected the spirit of the age. Many pagans (like Justin Martyr, for example) were attracted by the arguments of Christian apologists, who were willing to debate their intellectual opponents and who consciously appealed to educated pagans on philosophical grounds.[95] But although it seems reasonable to infer that we have in the writings of the apologists a better gauge of mainstream Christian opinion regarding miraculous healing than in the sensational world of second-century pseudepigrapha, we cannot discount its presence in Christian circles altogether.

The Third and Fourth Centuries

The number of Christians who sought religious healing for diseases in the third century appears to have remained small. A few fathers, like Cyprian (c. 200–258), claimed healing that could be brought about by means of exorcism and the sacraments, particularly baptism administered to those who were sick, although the evidence does not suggest that Cyprian routinely sought religious as opposed to medical healing.[96] But it was the fourth century that witnessed an increase in exorcism and miraculous forms of healing among Christians that reflected a credulity characteristic of the age, which was found among pagans and Christians alike. More instances of miraculous healing are reported from the fourth century than from the three preceding centuries combined. In part this might be explained by a greater fullness of sources. Some forms of Christian healing in the fourth century have traditionally been explained as the absorption into the church of popular manifestations of paganism after the legalization of Christianity by Constantine in 313. That explanation contains some truth,[97] but the issue is more complex, and scholars such as E. R. Dodds and Peter Brown have argued that a change in mental outlook came to characterize late antiquity, which was becoming increasingly focused on the supernatural.[98] The view that the second and third centuries witnessed the rise of a movement in society, which included intellectuals, away from rationalism and toward mysticism lacks sufficient evidence, however, to be convincing.[99] It is in the *fourth* century that one begins to see manifestations of a new outlook, particularly in the rise of Christian asceticism, which Peter Brown terms "the *leitmotiv* of the religious revolution of Late Antiquity."[100] Brown argues that the issue is not one of popular superstition or pagan survivals entering the church but of a struggle over control of the relics of the saints in which the church claimed control by making them public and revered rather than private, incorporating them into churches and within rather than outside the city walls, as they had been in pagan culture.

Asceticism (the practice of strict self-denial as a spiritual discipline) that arose in the late third century came to exercise a strong influence on Christianity in the fourth. New Testament writers had urged self-denial in the form of moral purity, detachment from the world, and rejection of its pleasures.[101] The asceticism that was introduced to Christianity in the late third century, however, went considerably beyond the pattern of the New Testament and the first three centuries of Christian practice. It idealized virginity and celibacy and preached contempt for the material world in general and for the body in particular. In its mildest form (among the Encratites) it involved sexual continence and abstinence from wine and meat. In its more extreme forms it held that only the spiritual world was good while the material world was evil and must be rejected. This dualism characterized some late Greek philosophies and religious groups (particularly the mystery religions) and can be found in some Jewish writings of the first century B.C. It was adopted in the second century after Christ by such Christian groups as the Gnostics, Manicheans, and Marcionites that the church branded heretical. Another kind of asceticism, adopted by some church fathers, such as Clement of Alexandria (c. 150–c. 215) and Origen, denigrated the body but developed a theological basis for asceticism that was not rooted in a dualistic rejection of the material world. Most fathers, however, rejected both the mild and extreme forms of asceticism, regarding the body as morally neutral and subservient to the soul in its warfare against sin.[102]

The ascetic outlook, with its denigration of the body, was one that was widely held by pagans in late antiquity, and it had a strong appeal to Christians.[103] Whereas earlier Christians had regarded suffering as a necessary part of life in this world, which God sometimes used for spiritual edification, most did not actively seek it. Many ascetics, however, sought suffering for expiatory or purificatory ends by subjecting their bodies to a variety of disciplines. The mortification of the flesh sometimes manifested itself in extreme ways. An early monk, Macarius (c. 300–c. 390), as penance for having killed a fly in anger, permitted poisonous flies to sting his naked flesh for six months. With the spread of monasticism in the fourth and fifth centuries, the influence of asceticism grew, but not all forms of monasticism placed the same emphasis on the mortification of the body. In general, Eastern monasticism (especially in its earlier anchoritic or solitary form) emphasized the denial of the body, while Western monasticism discouraged its rigorous forms in favor of a disciplined life that was characterized by practicality and charity. When in 313 Christianity became a legal religion and Christians no longer suffered persecution, ascetics over time replaced martyrs in the popular mind as the new spiritual heroes.[104] Because they sought by a daily regimen of self-denial to overcome the material world, ascetics enjoyed an exalted reputation, especially in the society of the Eastern Empire. They became the new spiritual elite, and their life of rigorous discipline came to be viewed by ordinary Christians as a daily martyrdom. From the mid-fourth century, most of the leaders of the church in both the East and West regarded asceticism as the path to spiritual perfection.

An overwhelming and sometimes uncritical acceptance of miracles of healing emerged in the latter half of the fourth century that continued throughout late antiquity.[105] It is not only the abundance of reported miracles that is striking but also their ubiquity among all classes of society. Nearly everyone seemed to be able to report cases of miraculous healing of which he or she had personal acquaintance.

The greatest preachers, scholars, and theologians of the age were enthusiastic in their acceptance of reputed miracles of healing, even of those whose credibility seems to the modern reader to be lacking.[106] Athanasius (c. 296–393), Ambrose, Jerome, John Chrysostom, and Augustine believed in the reality of miraculous healing as a contemporary phenomenon and encouraged the dissemination of miracle stories. Hence one can speak of not only a quantitative but also a qualitative change in this regard when one compares the late fourth century with the previous centuries of Christian history, in which reports of miracles are general, secondhand, and cautious.[107] The change can be seen strikingly in Augustine. Early in his Christian life Augustine accepted the opinion that miracles no longer occurred, having ended with the age of the apostles. This view is explicitly stated in his treatise *Of True Religion,* penned in 390, in which he writes that men no longer need miraculous proofs of their faith, which rely on the authority of scripture, since reason can now lead to understanding and knowledge of the truth and virtue.[108] Augustine, in fact, ridiculed claims of Donatist miracles.[109] Later in life he began to change his mind, particularly after the bones of the martyr Stephen were brought to Hippo in 424 and allegedly wrought some seventy miracles in less than two years.[110] He collected accounts of these and other healings and cataloged a large number of them in book 22 of *The City of God.* "Like most Late Antique men," writes Peter Brown, "Augustine was credulous without necessarily being superstitious," a statement that is amply demonstrated by the accounts of miracles that he included in *The City of God.*[111]

Rowan Greer argues that the emerging Christian interest in the miraculous was a new phenomenon that grew out of the Constantinian revolution in religion. Greer contrasts the perception of the first-century miracles described in the Gospels, as they were interpreted by the church fathers, with those of the fourth century, which he finds very different in intention. The accounts in the Gospels depict the miracles of Christ as having demonstrated his deity but not having produced faith in unbelievers. Their message was a theological one: they pointed to human redemption and to resurrection in the age to come.[112] Greer cites the homilies of fathers like Augustine and John Chrysostom to demonstrate that they did not consider the miracles of Christ to be particularly important in themselves. They were a phenomenon of the past, not the present.[113] Their value for the fourth century was chiefly a moral and homiletical one, while their role was one of Christianizing the traditional philosophical quest for personal virtue.[114] Augustine's homilies reflect his early view of miracles, in which he finds their primary importance in their spiritual or allegorical meaning.[115] He believes that miracles must be understood within their theological context and that they gain their significance by being rooted in the life of the church.[116] John Chrysostom, in a commonly held view of the fathers, argued that miracles had ceased with the apostolic age.[117] They were signs that pointed to Christ's triumph over Satan both in his death and in his power to indwell believers and enable them to live virtuous lives.[118]

With the recognition of Christianity in 313, argues Greer, Christians began to view the Roman Empire as a new Christian commonwealth that owed its existence to the kingdom of God, which was in the process of transforming the world. The result of this process was a gradual sacralization of the Roman Empire in the fourth century. The rise of the cult of saints marked an important element in this process for Christians, who saw what appeared to be the power of heaven being established on earth before their very eyes.[119] In contrast with their earlier lack of importance to fathers like Augustine and Chrysostom, miracles now became an important dimension of the Constantinian church, focused as they were in the West on relics of the saints and in the East on the presence of holy men. Greer attributes this change to the transformation of the social setting that provided the context for the new miracles. Holy men became benefactors and patrons, who performed special acts of God that assumed an importance in the account of the ecclesiastical historian Sozomen (fl. fifth century), for example, that they had not had for the earlier historian Eusebius (c. 260–c. 340). Greer considers this change a "shift of sensibility" away from the emphasis on God's general providential ordering of history that one finds in Eusebius to one of particular miraculous events that one sees in Sozomen's history of the church.[120]

In late antiquity, magic came increasingly to be used for healing by Christians and pagans alike. It has been argued that by the late third century the old Roman religious institutions had lost their appeal to all social classes. There was, as a consequence, a spiritual void that was filled by a variety of new religious manifestations, including the growing influence of magic, which was felt even in the highest intellectual circles.[121] Roman law from the earliest times strictly prohibited malicious magic (magic used to harm) and harshly punished its practitioners.[122] But benevolent magic, such as that which Cato the Elder (234–149 B.C.) had employed for the cure of sprains, was not condemned by law. The Theodosian Code (438) states, in a law regarding magic promulgated by Constantine, that "remedies sought for human bodies shall not be involved in criminal accusation."[123] Augustine and other fathers, however, considered dependence on magical powers and devices reprehensible because they attributed those powers to demonic forces.[124] For more than three centuries Christians had condemned the use of all magic, including charms and amulets.[125] Thus John Chrysostom commended in a sermon a mother who preferred to allow her sick child to die rather than to use amulets, even though she believed that such means would be effective and was urged by Christian friends to employ them.[126] But in the fourth century, as large numbers of nominal converts entered the church following its legalization and growing respectability, they brought pagan attitudes and practices, such as magic, with them.[127] It has been suggested that the supposed increase in the practice of magic in late antiquity indicates not so much a greater use as it does an increase in our sources for magic. But the fact that the fourth-century synods of Ancyra and Laodicea found it necessary to prohibit magic and threaten excommunication for priests who engage in magical practices argues that Christians had begun to adopt them widely.[128] Augustine complained of Christians who consulted astrologers after having unsuccessfully sought healing through prayer and natural remedies.[129] He wrote of Christian mothers who, in seeking healing for their children, used amulets and incantations and sometimes even offered sacrifices to the pagan gods in the hope of obtaining a cure.[130]

It was not always clear to Christians, however, what constituted magic. Augustine maintained that it was one thing to consume an herb for stomach pain and quite another to wear the herb around one's neck for the same purpose. He approved of the former practice, which he called a wholesome mixture; he condemned the latter as a superstitious charm. He conceded that wearing an herb around one's neck might be effective because of its natural virtue and thought it acceptable as long as incantations and magical symbols were not used in conjunction with it.[131] The confusion regarding what constituted magic is evident also in Augustine's indignation at Christians who mingled the name of Jesus with their incantations and in his ambivalence concerning mothers who saw baptism as a possible remedy for the healing of their sick children.[132] Augustine himself had, as a boy, begged his mother, Monica, to allow him to be baptized when he was sick, not only for the sake of his soul but for physical healing as well.[133] The strong repudiation of magical practices by Christians in the fourth century indicates both a religious confrontation with pagan practices that was undertaken to define Christian belief and an attempt to prevent their widespread adoption by Christians. It indicates that these practices (such as the use of amulets) were commonly employed by Christians, some of whom may not have viewed them as either magic or specifically pagan. In some cases the church provided alternatives, such as the sign of the cross, which was a magic more powerful than amulets.[134]

In the middle of the fourth century there was a pronounced increase in the number of Christian miracles reported and in their sensational and sometimes magical character.[135] The major source of this phenomenon is likely to have been desert fathers in the Eastern Empire, such as Anthony (251?–356) and his disciple Pachomius (c. 290–346), who came to exercise widespread influence and whom popular legend credited with many miracles. The classical world had often sought healing from seers and shamans: the *iatromanteis* of archaic Greece or the ascetic wandering teachers of the late first and second centuries after Christ who claimed to possess miraculous powers. In a sense, the Christian ascetics of the fourth and fifth centuries were descendants of these classical models. Miracle workers had appeared from time to time in the early centuries of Christianity, but they were usually the founders of new and often heterodox Christian sects, whose miracles were attributed by the orthodox to demonic powers. The desert fathers, however, were generally orthodox (as the contemporary church defined orthodoxy), and as their reputation grew, some of them were sought out by ordinary Christians for spiritual counsel and physical healing. Peter Brown attributes their appeal to the feeling of security, leadership, and personal warmth that they offered to a disintegrating and increasingly impersonal society. Be that as it may, what accounts for their rise in late antiquity is the change in atmosphere that made their appeal possible. Their ability to heal was attributed to the control that ascetics possessed over their own body through discipline and mortification.[136]

Athanasius wrote in Greek a life of Anthony, the founder of anchoritic monasticism, shortly after the latter's death in 356, and it was soon translated into Latin, creating a new genre of literature known as hagiography. Lives of saints, such as Gregory of Nyssa's lives of Gregory Thaumaturgus and Saint Macrina, proliferated, inspired by the enormous popularity of Athanasius's work, and they came to constitute the most popular form of Christian literature in the late fourth and fifth centuries. These lives described the miraculous exploits that had come to be attributed to the ascetics: their casting out of demons, miraculous healing of diseases, and raising from the dead. The ascetics were said to effect miraculous cures by prayer, making the sign of the cross, laying their hands on the afflicted, or applying bread, oil, water, or garments that they had blessed. Typical of these lives was Jerome's life of Hilarion (c. 291–371), a disciple of Anthony, whom Jerome credited with having performed many miracles of healing. They included restoring sight to a woman who had been blind for ten years, curing paralysis and dropsy, and casting out demons from those who had been possessed (including a possessed camel who had been responsible for many deaths).[137] Gregory of Nyssa's life of Gregory Thaumaturgus, the "Wonder-worker" (c. 213–c. 270), similarly recounts many miracles of healing that were attributed to a popular bishop in Pontus.[138] Stories abounded of every kind of physical disability that was healed by ascetics: leprosy, madness, paralysis, the loss of fingers, and severe wounds. Miracles of healing were also attributed to bishops like Ambrose, the influential bishop of Milan. Many of the healings involved the use of what can only be called Christian magic. For some Christians the name of Jesus became an irresistible spell and the sign of the cross an all-powerful charm.[139]

With the veneration of ascetics who could be looked to for healing came a new interest in martyrs and relics (the material remains of saints or objects that had some contact with them).[140] The remains of the earliest Christian martyrs had been venerated because the martyrs were thought to have been especially blessed of God, since they had proven their faith by their willingness to die for it. Hence their tombs were honored and they attracted pilgrims, who began to attribute miracles and cures to them. Miracle-working power was believed to reside not only in the bones of the martyrs and other holy persons but also in their garments and objects with which they had been associated. The relics of saints or martyrs extended to posterity the benefits that the saints had conferred on those in need during their own lifetime. The large number of converts from paganism after the legalization of Christianity brought into the church a reverence for relics that was a popular feature of pagan cults. Their veneration had begun as early as the second century, but from the mid-fourth century there was a rapid increase in the quest for relics and the building of shrines, which were accompanied by numerous alleged healings and purported manifestations of demonic activity. The tombs of martyrs at first challenged and later replaced pagan hero cults during the fourth century, as martyrs' shrines came to be celebrated for the cures they produced. Even some pagan healing shrines were eventually taken over by Christians.[141] "Like the old gods," writes A. H. M. Jones,

> they cured the sick, gave children to barren women, protected travelers from perils of sea and land, detected perjurers and foretold the future. Some acquired widespread fame for special power. SS. Cyrus and John, the physicians who charged no fee, were celebrated for their cures, and their shrine at Canopus, near Alexandria, was thronged by sufferers from all the provinces, as in the old days had been the temple of Asclepius at Aegae. But the main function of the saints and martyrs in the popular religion of the day was to replace the old gods as local patrons and protectors.[142]

Peter Brown has observed that in late antiquity it was in the Eastern Roman Empire that holy men enjoyed popularity, whereas in the West their place was taken by the cult of relics and tombs. He argues that the increasing claims of the hierarchical structure of Western Christianity, centered on the bishops of Rome, prevented the emergence of holy men in the West, while in the East they were allowed to flourish because of a lack of similar power claimed or exercised by the patriarch of Constantinople. But Rowan Greer maintains that what was novel about the cult of tombs and relics was their appropriation and organization by the church that gave them a central place in the Christian community.[143] Miracles were thereby given an official role in supporting the new Christian commonwealth. No longer did dead saints compete in authority with the hierarchy; they were under its control, and their own authority had been tamed and domesticated.[144] Not so with living saints, who were still able to challenge the authority of the church. Greer views the development of the cult of the saints as setting the stage for the Western Middle Ages.[145]

There is no question that a heightened supernaturalism came to characterize late antiquity in the fourth and fifth centuries. But the past generation of scholars has overemphasized the influence of demons, magic, and miracles on the thought and practice of Christians of late antiquity. No one has stressed this influence more than Ramsay MacMullen, who writes that early fourth-century attitudes among Christians (as among pagans) were marked by a nearly universal focus on signs and wonders, the ubiquity of demons and supernatural interventions, and frequent recourse to magic.[146] "But as a darkness of irrationality thickened over the declining centuries of the Roman empire, superstition blacked out the clearer lights of religion, wizards masqueraded as philosophers, and the fears of the masses took hold on those who passed for educated and enlightened."[147] The Gibbonesque language betrays the author's caricature. The exaggerated contrast that has often been drawn, by MacMullen and others,[148] between the credulous age of miracles and a modern understanding of the nature of reality is nicely illustrated by Harold Remus.[149] While some Christian literature of the fourth century, particular saints' lives and apocryphal acts, are indeed marked by an exaggerated supernaturalism, much of it is not. Rational approaches to religion continued to exercise a predominant influence on the thinking of ordinary Christians.[150] MacMullen's assertion that miracles succeeded in offering a more persuasive appeal than did preaching and apologetics is unsubstantiated.[151] One only has to point out that claims of miraculous healing are not responsible for most of the celebrated conversions to Christianity of the fourth century.[152] Christians did not offer the same promise of healing to pagans that the temple healing of Asclepius could.[153]

The most persuasive argument against the thesis that Christians helped to create a mentality that was marked by the ubiquity of miracles and magic, in which they attracted proselytes by their success in miraculous healing, is that in spite of the appeal of magical charms and relics in the West in the late fourth century, as well as the popularity of ascetics in the Eastern Empire, there appears to have been no diminution in Christians' seeking healing from physicians. While I shall reserve arguments in support of this assertion for a later chapter, it will suffice at this point merely to observe that the earliest hospitals began to be established by Christians in cities throughout the Eastern Empire at the same time that miraculous claims of healing were making their appeal—in the latter half of the fourth century—and that these hospitals were in some cases staffed by physicians and attendants. Even ascetics, as we shall see, were by no means averse to recommending the use of medicine when they believed it would be efficacious, though they were sparing in using it themselves. Miraculous healing did not replace for Christians their ordinary reliance on medicine. As Anne Merideth reminds us, "Hagiographical literature, in particular, celebrates the extraordinary and the miraculous rather than the ordinary and mundane. When immersed in such texts, it is all too easy to assume that ritual healings and miraculous cures dominated the daily existence of ancient Christians."[154] There is, moreover, an important inference to be drawn from the nature of these sources. Miraculous healing, as it became a highly visible phenomenon in the late fourth century, was derivative of the ascetic movement. Its source was not ritual healing administered within the context of the liturgy or practice of the church. It was, rather, a highly visible manifestation of divine power that only holy men could exercise and that had not previously been seen in the same way in the church.[155] If Rowan Greer is correct (as I believe he is) in arguing that the nature of miracles in the fourth century is novel—of a different character altogether than earlier miracles of the sort described in the Gospels—and that they owe their existence to a change in sensibility made possible by the legalization of Christianity, then it is not surprising that we see an explosion of claims of miraculous healing in late antiquity. The frequency of miracles of healing owed its existence to the new role that holy men had acquired in the new Christian commonwealth under Constantine. In the West relics, not ascetics, served as vehicles of healing. Both were important aspects of the sacralization of Roman society that followed Christianity's recognition, at a time when a very different cultural context not only made possible a new kind of miraculous healing but created a distinctive role for it.

Conclusion

The New Testament depicts Jesus's miracles of healing as signs of the coming of God's kingdom, the external manifestation of the supernatural in the natural world, rather than as normative models of physical healing intended for the Christian community. The New Testament Epistles indicate that Christians experienced ordinary illnesses, of which they were sometimes healed and sometimes not healed. Biblical writers, moreover, do not condemn medicine. And outside the Gospels and Acts there are comparatively few references to miraculous healing through the second century. The writings of the Apostolic Fathers in the first half of the second century are devoid of reference to miraculous healing. The apologetic literature of the latter half speaks of it, but in a vague and general way that lacks specific examples, which makes it questionable that contemporary events are being alluded to. One finds, moreover, little attempt to exploit for apologetic purposes specific instances of Christian healing in an age in which testimonies to miraculous cures by pagan gods were common.

Although there is some evidence of claims to miraculous healing in Christian communities of the third century, a dramatic explosion of accounts of healing occurred in the late fourth century, as Christians increasingly sought miraculous cures. This is related to the popularity of the holy man in late antiquity in the Eastern Roman Empire and the heightened appeal of miraculous healing in both East and West. Healing came to be sought by Christians through a variety of means: invocation of the name of Christ, prayer and fasting, the sign of the cross, the imposition of hands, unction, the use of amulets, and exorcism.[156] It was acquired through the agency of ascetics, saints' relics, bishops, and others. Christians began to advertise their miracle cures, as the pagans had long done. After adopting the new view, Augustine rebuked Christians who did not publicize their miraculous healings, which he believed ought to be more widely known.[157] We see, too, the rise of Christian aretalogies, found chiefly in the lives of saints, which enjoyed great vogue in the late fourth and fifth centuries and prepared for the vast popularity of Athanasius's life of Anthony. The new resort to miraculous forms of healing in late antiquity did not, however, as we shall see in the next two chapters, lead to a decreasing reliance on medicine by Christians. Probably the majority of Christians continued to seek out physicians or employ home or traditional remedies, while the establishment of hospitals extended medical care to the indigent, particularly to the urban homeless who were previously without the means to obtain it. Christianity was never a religion of healing in the sense that Harnack described it, comparable to the great healing religions of Asclepius and Serapis. At no period was healing central to the early Christian message, and it always remained peripheral to a gospel that offered reconciliation to God and eternal salvation to sinners.

The Basis of Christian Medical Philanthropy

Christianity spread rapidly in the first century, owing to its extensive missionary activity, from its birthplace in Palestine throughout the Roman Empire. By about A.D. 60 the new faith had been carried to most parts of the eastern Mediterranean and as far west as Rome. In A.D. 64 Nero accused the Christian community in Rome of having set fire to the city, and in order to divert suspicion from himself, he began actively to prosecute them. Thereafter, for the next 250 years Christians faced sporadic persecution. Roman officials regarded them as traitors for their refusal to offer sacrifice to the emperor as a god, and as atheists for their failure to participate in public pagan worship.[1] Yet by the middle of the second century Christian communities thrived in most major and many minor cities of the Roman Empire.[2] At the same time that they were undergoing persecution, Christians carried out an active program of philanthropy, which included the widespread care of the sick both within their own community and, especially during times of plague, outside of it. Their long experience in medical charity prepared the way for the eventual establishment of the first hospitals as specifically Christian institutions, which followed the legalization of Christianity by Constantine in 313 by a half century. The conceptual and ideological origins of Christian medical philanthropy were rooted in a very different set of values than was the concept of beneficence in the classical world. We shall explore both in this chapter.

Medical Philanthropy in the Graeco-Roman World

The term *philanthropy* is derived from the Greek word *philanthropia*, which means "a love of mankind."[3] The original meaning of the word was the benevolence of the gods for humans, a concern that manifested itself in the granting of gifts and benefits. By a natural extension of meaning the word came to refer as well to the munificence and generosity of rulers toward their subjects and to the friendly relations between citizens and states. In all these meanings we find common elements of condescension and the giving of gifts or benefits that the word never lost. In the fourth century B.C. the word came to be used with the more general meaning of "kindly, friendly, genial" in reference to personal and social relationships. It is widely used in this sense—for example, in the Hippocratic Corpus—to indicate a kindliness, courtesy, and decent feeling toward others. It is doubtless with this meaning in mind that Diogenes Laertius (fl. first half of the third century after Christ) writes that *philanthropia* may take three forms: that of salutation, of assisting one in distress, and of fondness for giving dinners. "Thus philanthropy is shown either by a courteous address, or by conferring benefits, or by hospitality and the promotion of social intercourse."[4] In the Hellenistic period the word took on a much more comprehensive meaning and was sometimes used to express a love of humanity, suggesting a general feeling of concern for the well-being of one's fellows. Yet even in this sense *philanthropia* continued to retain its original meaning of a relationship between a social superior and an inferior, a condescending benevolence that reflected the limitation in the classical world of the philanthropic impulse.[5]

In general it may be said that philanthropy among the Greeks did not take the form of private charity or of a personal concern for those in need, such as orphans, widows, or the sick.[6] There was no religious or ethical impulse for almsgiving; nor was pity recognized either as a desirable emotional response to need and suffering or as a motive for charity. In contrast with the emphasis in Judaism on God as particularly concerned for the welfare of the poor,[7] the Greek and Roman gods showed little pity on them; indeed, they showed greater regard for the powerful, who could offer them sacrifices. Pity as an emotion was reserved not for the indigent but for those—mostly members of the upper classes—who had experienced a reversal of fortune that had reduced them to poverty; because the lower classes had never experienced a catastrophic fall, they could not deserve pity. As a motive for assisting those in need, pity was shown by those who, on the one hand, could sympathize with members of their own class in need and, on the other, might hope to build up a fund of goodwill in case they should experience a similar misfortune. The Stoics regarded pity "not as a liberating emotion necessary to inspire the selfless service of others, but as an emotion which enslaved a man's mind and spirit, and undermined the good man's claim to self-sufficiency and self-command."[8] The basis of generosity or of any moral action for a Stoic should be rational rather than emotional; the latter was regarded as impulsive and subjective, the former as objective, universal, and humanitarian. This was typically the classical view.

It was only on a quid pro quo basis that pity might serve as a motive for giving. Givers hoped that, if they ever found themselves in need, they would receive pity and aid, since they had earned pity by displaying it themselves. Hence pity might most properly be felt for the members of one's own class, from whom reciprocation could be expected. When it was shown more generally, it was out of an instinctive sympathy for the human condition, as in the *arai Bouzygeiai*, which were "curses which were called down upon any man who failed to provide water for the thirsty, fire for anyone in need of it, burial for an unburied corpse, or directions for a lost traveler."[9] The motivation for such acts is found in the statement attributed to Aristotle, "I gave not to the man, but to mankind."[10] Here benevolence is a form of hospitality rather than of justice or moral or religious obligation. One is to do as one would be done by. Even in such simple acts of human concern, there was an eye to reciprocity. One might someday require similar assistance, perhaps even from the person one had helped in the past.

"For where there is love of man [*philanthropia*]," reads a well-known passage in the Hippocratic treatise *Precepts* (6), "there is also love of the art [*philotechnia*]."[11] This statement appears in the middle of a section dealing with the question of medical fees. The opening sentence begins with the admonition "I urge you not to be too unkind." The word translated "unkind" is the noun *apanthropia*, which is an antonym for *philanthropia*.[12] Hence the physician is urged not to be too "unphilanthropic" but to consider his patients' financial means and to treat gratuitously the stranger in financial straits. "For where there is love of man there is also love of the art" is often understood to mean that when the physician is a lover of people he will be, as a result, also a lover of his art. We should expect the dictum to be followed by the conclusion that if a physician is motivated by both *philanthropia* and his *philotechnia* he will extend compassionate care to his patients. But instead we find that the *philanthropia* belongs to the physician and the *philotechnia* to the patient: "For some patients, though conscious that their condition is perilous, recover their health simply through their contentment with the goodness [*epieikeia*, clemency, natural mildness] of the physician."[13] Hence the patient's response to the physician's *philanthropia* takes the form of *philotechnia*, love of the physician's art, which reveals the patient's contentment with the physician's *epieikeia* or kindness. This contentment greatly aids in the curative process.[14] The *philanthropia* of the physician here seems to denote an attitude of kindliness and charity.

A similar interpretation must be given to a passage in the Hippocratic work *On the Physician* (1).[15] Here the concern is with the proper deportment that a physician's dignity requires. In the context of much sage advice on medical etiquette and morality appears the statement that the physician "must be a gentleman" who is "grave and *philanthropos*." The last word is the adjective derived from *philanthropia.* Is this passage urging the Greek physician to be a "lover of mankind" as a motive for his practice? To this question an unequivocal "no" must be given. W. H. S. Jones nicely captures the meaning of *philanthropos* here when he translates it in this passage as "kind to all."[16] A few sentences later the physician is urged not to appear harsh, for then he would seem to be *misanthropos*, which here probably means little more than "unkind." In both passages *philanthropia* seems to designate "a proper behavior toward those with whom the physician comes in contact during treatment; it is viewed as a minor social virtue."[17] Hence it can be little more than a guide, not an impulse or motivation, for the practice of medicine. Rather, the motivation of the classical physician to practice medicine seems more often to have been *philotimia* ("love of honor") than *philanthropia.* This exchange of giving, which is found as early as Homer, had its origin in an aristocratic society of equals for whom giving and countergiving cemented friendships. Such a relationship, which brought advantages to both parties, eventually spread beyond the upper classes and came to involve, to some degree, all members of the community. It was, for example, the basis of the patron-client relationship in Rome, where it involved services (*beneficia*) rather than merely material gifts. It came as well to include the relationship between unequals, as we find in the word *philanthropia.* Where this association existed between the wealthy and the poor, the only return that could be made by the poor was "honor" in the form of public or private recognition of the philanthropy of the benefactor. The desire for honor and public recognition served as one of the chief motives of personal behavior in the classical world. Public philanthropy was one of the most important means of obtaining it in the community, and it was by no means uncommon for benefactors to admit that they were giving in return for public recognition. "The Greeks, in particular, believed that the good man would pursue honor, admiring as they did a strong competitive element in man's psychology."[18] "It is quite clear," writes Cicero, "that most people are generous in their gifts not so much by natural inclination as by reason of the lure of honor—they simply want to be seen as beneficent."[19] *Euergesia*, a good service or benefit that was well publicized and bestowed in the expectation of increasing one's personal reputation, was a characteristic civic ideal that enjoyed remarkable longevity in classical society.[20]

The return of honor for a benefaction had special reference to the operation of the city-state in the Graeco-Roman world, where many of the financial burdens of the community were met by the wealthy class either by the holding of a liturgy (a public office that often required considerable personal expense) or by an appeal to the wealthy for a public subscription. It was a regular practice to obtain a portion of the public revenue of a city from the gifts of the wealthy; in the case of a subscription a motion would be made to establish a fund for a need, to which the wealthy members of the community were expected to contribute. The impulse for such giving was, positively, *philotimia* ("love of honor") or *philodoxia* ("love of glory") and, negatively, the threat that the wealthy would be exposed to prosecutions that might result in the loss of either their position of honor or their wealth. As a return for their subscription, the community often rewarded wealthy benefactors by setting up honorary inscriptions, which recorded on stone or bronze, sometimes in great detail, the nature and amount of the benefaction. Thousands of these inscriptions remain today that testify to the public philanthropy of the wealthy—and others, such as physicians, teachers, and philosophers—who made public benefactions or performed some public service.

The impulse for giving was not pity. "Broadly speaking, pity for the poor had little place in the normal Greek character, and consequently for the poor, as such, no provision usually existed; the idea of democracy and equality was so strong that anything done must be done for all alike; there was nothing corresponding to our mass or privately organized charities and hospitals."[21] Hence, when gifts were made or services performed, they were intended for the entire community. No distinction was made between the destitute and others.[22] It is in this context that the numerous public inscriptions honoring physicians must be understood. "Nothing leaves a more pleasing impression," write Tarn and Griffith, "than the numerous decrees of thanks passed to physicians."[23] The physicians so honored are described as tireless in their services on behalf of the community, devoted to their profession, making themselves available to all who need their services, serving rich and poor, citizen and foreigner alike, remitting fees, remaining in the city during an epidemic. Here, if anywhere, there seems to be prima facie evidence of a genuinely disinterested "love of mankind" as a motive for medical care. Yet there is nothing to distinguish honorary decrees for physicians from the whole class of honorary decrees passed by Greek cities for benefactors of all kinds. The language is formulaic, and the benefactions for which physicians are honored can be at least partially paralleled elsewhere. Thus if physicians sometimes remitted fees for those unable to pay, so did philosophers on occasion, and they too were honored by public decrees.[24] Physicians were rewarded for their service to the community in the ordinary way in which communities rewarded benefactors: by public honors voted them. For the physician, according to *Precepts,* money is of secondary importance to honor: "The quickness of the disease . . . spurs on the good doctor not to seek his profit, but rather to lay hold on reputation."[25] The honorary inscriptions suggest that *philotimia* was an important, if not the chief, motivation of many classical physicians.[26]

Henry Sigerist writes that "every period has an ideal physician in mind, indeed must have one."[27] Irrespective of how far short of the ideal many physicians may have fallen, an ideal did exist. Was a physician considered to be a physician only insofar as he lived up to such an ideal? Naturally most people, laymen and physicians alike, never asked the question, but some writers did. Plato, for example, in the *Republic* (340C–347A) discusses the question of whether self-interest is the motive behind all human efforts, especially political activity. A comparison is made between politics and various arts, including medicine. In this context the question is asked whether the physician qua physician is a healer or an earner. Qua physician he is exclusively a healer, since in that capacity his interest is entirely in providing the advantage for which his art exists. Acting qua physician, he does not seek his own or his art's advantage but only his patient's. His earning of an income or his gaining of honor from his art is itself a subsidiary art and follows from the practice of his primary art. Hence the motivation for practicing any art, whether it be for money or honor, is quite irrelevant to the integrity of the art itself, since the raison d'être of the art is the furnishing of the good for which it was created. Motivation is not the issue; competence in the art is what is essential, for without competence the putative practitioner of any art fails in fact to be a practitioner, owing to his incompetence to achieve those ends for which his art exists.

Galen, in a work entitled *On the Doctrines of Hippocrates and Plato,*[28] discusses this specific passage from the *Republic.* After summarizing the argument, he writes, "Some practice the medical art for monetary gain, some because of the exemptions granted them by the laws, some from love of their fellow men [*dia philanthropian*], others again for the fame and honor that attend the profession."[29] Galen goes on to say that they are all called physicians, insofar as they provide health, but insofar as they are led by different motives, "one will be called a *philanthropos,* another a lover of honor, another of fame, still another a money-maker." Therefore the aim of physicians qua physicians is neither glory nor reward, as the Empiricist Menodotus wrote.[30] "This is the goal for Menodotus, but not for Diocles,[31] and not for Hippocrates and Empedocles[32] either, or for many other ancients, who treated men for *philanthropia*" (IX 5.6). Galen's understanding of the force of the words *philanthropia* and *philanthropos* seems to go beyond the spirit of the authors of the Hippocratic treatises *Precepts* and *On the Physician* even if he is not wholly different from them. The reason for this may lie in the influence of humanitarian and cosmopolitan ideas on both philosophical and popular ethics in Hellenistic and Roman thought. As has already been observed, after the fourth century B.C. the word *philanthropia* came to be used to express a comprehensive love of humankind and a common feeling of humanity. It has been suggested that this change was due to the growing cosmopolitanism that followed Alexander's conquest of the East or that it was the inevitable result of the lessening importance of the polis and the growing individualism of the fourth century.[33] In any event, the theme of a common kinship of humankind was taken up by Cynics and Stoics in Hellenistic Greece and the early Roman Empire.[34] One finds in the writings of the Stoics, particularly in Musonius Rufus, Seneca, Epictetus, and Marcus Aurelius, an emphasis on the brotherhood of all men, a love of one's enemy, and forgiveness of those who have done wrong to us. Philosophy—not religion—was regarded by the educated as the moral instructor of humanity, and it is apparent that cosmopolitan ideas of Stoicism influenced Roman society in, for example, the increasing amelioration of slavery. Stoic emphasis on sympathy and brotherhood seems to have influenced the concept of *philanthropia*, which was used to denote benevolent and civilized feeling toward all classes in the Roman Empire in the sense in which Galen seems to use the term.[35] Aulus Gellius reflects this meaning when he says that the Latin word *humanitas* is commonly taken to have the same meaning as the Greek word *philanthropia*, which signifies "a kind of friendly spirit and good-feeling towards all men without distinction."[36] Edelstein believes that "philanthropy became integrated into the ethical teaching of the dogmatic physicians not long before Galen's time, if indeed it was not Galen himself who accepted the ideal of philanthropy in accordance with his Stoic leanings."[37] There can be little doubt that Galen reflects the greater humanitarianism that was taught by the Stoics of his day.

While there is no exact equivalent of *philanthropia* in Latin, *humanitas* is a word that came to have many of the same associations.[38] Aulus Gellius, in the passage cited above, writes that the word has the force of the Greek *paideia*, "what we call education and training in the liberal arts." *Humanitas* comprehends the humane virtues that one expects of an educated person: politeness, tolerance, command of the social graces, but also kindliness, mercy, consideration of others. The word was a favorite of Cicero's, who defined those qualities that he believed a liberal education should produce in a person. It is not surprising that by the time of Gellius *humanitas* had come to be synonymous with *philanthropia*. It has been suggested that the Roman concept of *humanitas* goes even beyond the Greek concept of *philanthropia*, in that "it may have conveyed the idea of a warm, human sympathy for the weak and helpless in a measure which *philanthropia* never did."[39] This is an unlikely view, for the concept of *humanitas* was limited to a narrow circle of urban and educated aristocrats. Nevertheless, the word reflected the qualities that Romans expected to characterize their ruling class and that motivated much of the humane legislation of the early empire. Moreover, it has been argued that the concept of *humanitas* provided a distinctively Roman voice that went beyond *philanthropia* in its approach to issues of medical ethics.

Of all Galen's works there is perhaps none that is more fundamental for one's understanding of him than the short treatise entitled *That the Best Physician Is Also a Philosopher* (*Quod optimus medicus sit quoque philosophus*).[40] It is Galen's foundational principle that both medical research and treatment must be based on philosophy and therefore that the best physician must also be a philosopher. Galen did not limit the role of philosophy in medicine simply to supplying the scientific framework that natural philosophy provides but insisted that philosophy provide the ethical principles for medical theory and practice as well. Hence the best physician must be a philosopher, and as such he must be "self-controlled and just and immune to the temptations of pleasure and money; he must embody all the different characteristics of the moral life which are by their very nature interdependent."[41] A predominant feature of this moral life for Galen was *philanthropia.* This *philanthropia* manifested itself in his claim that he never demanded remuneration from any of his pupils or patients. Galen further tells us that he often provided for the various needs of his poorer patients.[42] In doing this he believed that he was following the example of the ancients, particularly of Hippocrates. Galen idealized Hippocrates and considered him an exemplar of medical probity and virtue. In this treatise Galen writes that Hippocrates declined the lucrative position of physician to a powerful Persian satrap in order to stay in Greece and take care of the poor. A little earlier in the same work, Galen calls medicine "an especially philanthropic art."[43] Hippocrates' spurning money and choosing to treat the poor freely are not the only proofs of his philanthropy that Galen cites to show that medicine is particularly philanthropic.[44] He adduces as additional evidence the facts that Hippocrates traveled about "to verify by experience what reasoning had already taught him about the nature of localities and waters"[45] and that he published his medical knowledge for the good of humankind. Galen admired Hippocrates as the model of the philosophic physician but ignored in his own writings the deontological treatises of the Hippocratic Corpus. Competence was Galen's overriding concern, and he believed that philosophy provided a necessary and sufficient basis for the physician's ethical behavior quite apart from reliance on an oath or specific Hippocratic treatises.[46]

For Galen medicine was an especially philanthropic art for two major reasons. First, regardless of whether or not *philanthropia* provided the motive for any particular physician, the art itself, when practiced by a competent physician, relieved humankind's sufferings. Second, if the physician was motivated by *philanthropia,* which he would be if he was also a philosopher, he would demonstrate his philanthropy in the ways in which both Hippocrates and Galen did: in the compassionate care of the destitute, the advancement of medical knowledge, and the dissemination of that knowledge to both contemporaries and posterity. In Owsei Temkin's words, philanthropy was for Galen "the love of mankind . . . and the concern for its future. . . . Galen's philanthropy is not only that of the physician, but more comprehensively that of a philosopher who subjectively delights in study and objectively labors for the good of mankind."[47]

Narrower in scope but deeper in its emphasis on compassion than Galen's view of philanthropy were the ideas espoused by Scribonius Largus, a Roman physician who lived in the first century of the Christian era. Scribonius composed a Latin treatise on drug recipes, to which he added a preface on the *professio* of medicine.[48] The preface does not deal with the profession or calling of the physician but refers to a public declaration by which the physician takes on the duties of medicine.[49] The term *professio* in Scribonius's time was a word charged with Stoic overtones.[50] Panaetius, a Stoic philosopher of the second century B.C. whose influence on Cicero permeates the latter's *On Duties (De officiis),* maintained that any legitimate role that one assumed had various duties (*officia*) that were central to it. If one who occupies a particular role is faithful to the *officia* inherent in that role, one is acting morally and justly in accordance with one's public declaration. If one is unfaithful to the *officia* of one's role, one is not only acting immorally but is violating the integrity of that declaration. When violating the *officia* of one's *professio,* one ceases to occupy that role and is no longer, for example, a judge, a lawyer, or a physician.

A proponent of drug therapy, Largus opposed those physicians who rejected drugs and employed only dietetics. He writes that some physicians who reject drugs do so out of ignorance, which is reprehensible, while others, who know the usefulness of drug therapy, deny it to their patients out of jealousy toward their colleagues who treat their patients more effectively. Such physicians are even more to be condemned than those who are simply ignorant. "They ought to be despised by gods and men, all those physicians whose heart is not full of compassion [*misericordia*] and humaneness [*humanitas*] consonant with the will [*voluntas*] of the *professio* itself." Therefore the physician will harm no one, and "because medicine does not have regard for men's circumstances or their character, she *[medicina]* will promise her succor equally to all who seek her help and she promises never to harm anyone." Citing the Hippocratic Oath (he is the first to mention it), he asserts that Hippocrates, in forbidding the practice of abortion, "had come a long way in the direction of preparing the hearts of his students for *humanitas.*" Medicine "is the science of helping, not of harming. Unless she strives in every way to succor the afflicted, she fails to provide for men the compassion [*misericordia*] that she promises."[51] We see in Largus's statements two essential features of the physician: (1) He must be competent; and (2) he must be motivated by compassion and humaneness, that is, he must be a "lover of mankind" in the sense in which *philanthropia* and *humanitas* were popularly used in his time. Failure to act in a compassionate and humane manner renders one no longer a physician. Largus's attitude is starkly different from the nearly negligible role that humanitarianism plays in the Hippocratic Corpus and indeed in previous medical ethics. And it differs from Galen's thought: Galen views competence as the only *essential* attribute of the physician, although the best physician is also a philosopher, and such a physician should, as a consequence, also be a "lover of mankind." *Philanthropia* was, for Galen, highly desirable but not essential for a physician. Not so in the case of Largus, for whom it was an essential feature of the true physician. So central were compassion and humaneness to Scribonius Largus that Kudlien has felt that he was "nearly Christian."[52] But he was not a Christian; he was a pagan who was significantly influenced by Stoic ideas that bear a prima facie resemblance to some Christian principles but went beyond those ideas to develop a distinctively Roman advance in humane medicine.

The Classical Basis for Human Worth

The question of whether the classical world possessed a religious or philosophical basis for a definition of the worth of human life (*dignitas humana*) that applied to all humankind has been widely debated. Two studies have concluded that this idea cannot be found in classical Greek and Roman authors.[53] The classical world believed in the dignity of humans, but it was in the dignity of the virtuous person, the person who possessed *arete* (excellence, virtue). This understanding was based on the belief that only a balanced and controlled personality that exhibited the recognized virtues of society could be deemed virtuous.[54] The Roman concept of *humanitas* was used to describe the humane virtues that were expected to be possessed by educated people, but they were virtues that were thought to characterize only a small group that belonged to the upper class of Roman society.[55] Human worth, then, was not regarded in the classical world as intrinsic. Nor did there exist any concept of inherent human rights.[56] Rights were defined judicially, and they depended on membership in a society (a family, kinship group, or state) that granted them. Those who lay outside (e.g., foreigners, slaves, foundlings) had no claim to any inherent rights, though they might be granted certain privileges, as were foreigners and slaves on occasion. Inequality was deemed a natural feature of life in the classical world, and it did not cause surprise or regret.[57] There was, for example, little sympathy in early Greek literature for the physically impaired or oppressed, an attitude that can be demonstrated to have characterized both popular and official opinion in virtually every period of classical antiquity.[58] Attitudes to the physically defective reflected the belief that health and physical wholeness were essential to human dignity, so much so that life without them was not thought to be worth living.[59] Citizenship, kinship, status, merit, and virtue formed the foundations of claims to the possession of human rights or human worth. Those who lacked them (e.g., orphans, slaves, foundlings, the physically defective, prisoners) had no claim to the rights that they alone guaranteed or even to a recognition of their human worth.[60]

The attribution of human identity to the unborn grew out of the question of whether an embryo was defined as a human being in its own right or merely an extension of the body of the mother. There existed no agreement among natural philosophers or medical writers regarding whether personhood began at conception, at birth, or at some point in between. A minority, specifically the Pythagoreans, believed that the fetus was animated from the time of conception. Another minority, particularly the Stoics, dated animation from the moment of birth. The majority of physicians and philosophers were gradualists, believing that animation developed in the fetus at one or another point during pregnancy. The pre-Socratics and Hippocratic writers debated how and when one could demarcate "formed" and "unformed" fetuses, and rival theories arose regarding animation, sense, and ensoulment. The debate over when human life begins was never resolved in antiquity, but it influenced not only subsequent classical thinking but also that of the Christian fathers. The biological data were inconclusive, and the medical issues were then, as now, not merely scientific ones but were intimately tied to philosophical assumptions.[61]

One might expect that the philosophical sects that arose in the Hellenistic age, and Stoicism in particular, would have by their teachings provided the foundation for the belief that all human beings are endowed with value and therefore possess basic rights.[62] Roman Stoicism was marked in the first two centuries of the Christian era by a cosmopolitanism and humanitarianism that affirmed the brotherhood of all human beings and the necessity of kindness (*beneficentia*) and humane treatment of every person, civilized or barbarian, slave or free, all of whom were regarded as possessing a divine spark.[63] Yet, as promising as this belief appears, it never developed into an explicit claim that all individuals possessed human rights, perhaps because the pantheistic theology of Stoicism prevented the uniqueness of the individual from being fully acknowledged.[64] There were, moreover, features of Stoic doctrine that were not hospitable to the development of the idea of an intrinsic human worth that applied to all persons. The Stoics cultivated an apathy to suffering because they believed that pain, sickness, and suffering were indifferent things (*adiaphora*).[65] Hence one finds a kind of hardness in Stoic teaching that has little place for the gentle virtues. As Lecky long ago observed, with particular reference to Stoicism, "friendship rather than love, hospitality rather than charity, magnanimity rather than tenderness, clemency rather than sympathy, are the characteristics of ancient goodness."[66] Although Stoicism aimed at a very high level of moral excellence, its practical influence was disappointing. Its suppression of the emotions and its elevated morality aimed too high for the ordinary individual, who could not hope to rise to the standard advocated by Stoics. As a result it had little influence on the morality of the masses. It is true that the influence of Stoicism on Roman law was extensive in ameliorating, for example, the treatment of slaves. But the Stoics' indifference to human suffering in general prevented them from actively seeking the protection of the weak. One finds in Stoicism, as in classical thought generally, a profound pessimism about human nature that led to a practical quietism. The Stoics were reluctant to attempt radical change in society or the amelioration of human institutions, believing that they were for the most part incapable of improvement.

The *Imago Dei* and Christian Principles of Philanthropy

Early Christian philanthropy was informed by the theological concept of the *imago Dei*, that humans were created in the image of God, a belief that was taken over from Judaism.[67] In Jewish religious practice Yahweh could not be visibly represented in any form (see Deut. 4:15–19). Hence Jews were prohibited from making images, which were characteristic of polytheistic worship. The nature of Yahweh was represented not by pictorial images but by the human race. Humans alone could be called the image of Yahweh because in their nature and being they reflected their Creator. The locus classicus of the concept of the *imago Dei* is Genesis 1:26–27: "Then God said, 'Let us make humankind in our image, according to our likeness; and let them have dominion over the fish of the sea, and over the birds of the air, and over the cattle, and over all the wild animals of the earth, and over every creeping thing that creeps upon the earth.' So God created humankind in his image, in the image of God he created them; male and female he created them." The belief that the image of God in humans had implications for the protection of human life in Judaism is suggested in Genesis 9:6, where Yahweh tells Noah, "Whoever sheds the blood of a human, by a human shall that person's blood be shed; for in his own image God made humankind." According to the Hebrew concept of the human personality, people were viewed as a unity rather than in dualistic terms. There were two elements in a person's nature, the "soul" (*nephesh*) and the "flesh" (*bāśār*). The soul was not made to exist apart from the flesh. To destroy the human body was to destroy the human personality, and thus it was an affront to the dignity of Yahweh, whose image (and therefore worth) humans bore. Hence in Hebrew thought human life possessed intrinsic value by virtue of its divine endowment in contrast to classical Graeco-Roman thought.[68] The concept of the *imago Dei* provided the basis for human value that was to become central to Jewish concepts of personhood. As a result features that were common to ancient society (child sacrifice, exposure of infants, infanticide, and emasculation) were not common in Israel.

The Hebrew concept of the *imago Dei* was carried over to the New Testament.[69] It is found without change, for example, in such passages as James 3:9 and 1 Corinthians 11:7. But the emphasis of the New Testament is soteriological and eschatological: it is concerned with the salvation and ultimate destiny of the fallen human race. In this regard the doctrine of the Incarnation is the major contribution of the New Testament to the concept of the *imago Dei*: "And the Word [*logos*] became flesh [*sarx*] and lived among us" (Jn. 1:14a). *Philanthropia* was not a word frequently used by early Christian writers.[70] It is found twice in the New Testament (Acts 28:2; Titus 3:4); in both instances it means "kindness."[71] The early Christians preferred a different word with a very different meaning: *agape*, a previously little-used and colorless word before it was given specific Christian content.[72] In the New Testament the concept of *agape* is rooted in the nature of God. "God is love [*agape*]," writes the apostle John (1 Jn. 4:8). It was God's love for humankind that brought about the Incarnation (Jn. 3:16). It was Christ's self-sacrificing love that led to his death on the cross as a ransom for humankind's redemption. And this love (*agape*) was expected to characterize those who professed his name. Hence any response to God was a response to his prevenient love: "We love because first he loved us" (1 Jn. 4:19). *Agape* was unlimited, freely given, sacrificial, and not dependent on the character of its object.

The Christian understanding of the *imago Dei*, viewed in the light of the doctrine of the Incarnation, was to have four important consequences for practical ethics that became increasingly apparent as Christianity began to penetrate the world of the Roman Empire. Together they represent a radical departure from the social ethics of classical paganism. The first was the impetus that the doctrine gave to Christian charity and philanthropy.[73] The classical world had no religious or ethical impulse for individual charity.[74] Personal concern for the poor and needy was an important theme in the Hebrew scriptures, which gave rise to the insistence in later Judaism (e.g., in the Apocrypha and the Talmud) that almsgiving is a duty and even the highest virtue.[75] This emphasis was appropriated by Christianity and is mentioned often in the pages of the New Testament, where charity is represented as an outgrowth of *agape*, which is rooted in the nature of God. Just as God loved humans, so they were expected to respond to divine love by extending love to a brother, who bore the image of God (Jn. 13:34–35). Love of God and devotion to Christ provided the motivation for love of others that had its practical outworking in charity (Mt. 25:34–40). Compassion was regarded as a manifestation of Christian love (Col. 3:12; 1 Jn. 3:17) and an essential element of the Christian's obligation to all people. This is succinctly expressed in the Clementine Homilies, which were written sometime before 380:

> Ye are the image of the invisible God. Whence let not those who would be pious say that idols are images of God, and therefore that it is right to worship them. For the image of God is man. He who wishes to be pious towards God does good to man, because the body of man bears the image of God. But all do not as yet bear his likeness, but the pure mind of the good soul does. However, as we know that man was made after the image and after the likeness of God, we tell you to be pious towards him, that the favour may be accounted as done to God, whose image he is. Therefore it behooves you to give honour to the image of God, which is man—in this wise: food to the hungry, drink to the thirsty, clothing to the naked, care to the sick, shelter to the stranger, and visiting him who is in prison, to help him as you can. And not to speak at length, whatever good things any one wishes for himself, so let him afford to another in need, and then a good reward can be reckoned to him as being pious towards the image of God. And by like reason, if he will not undertake to do these things, he shall be punished as neglecting the image [6].[76]

As we have seen, the classical concept of *philanthropia* was not merely insufficient to provide the motivation for private charity; it actively discouraged it.[77] In the Graeco-Roman world beneficence took the form of civic philanthropy on behalf of the community at large. Christianity, however, insisted that the love of God required the spontaneous manifestation of personal charity toward one's brothers: one could not claim to love God without loving his brother (1 Jn. 4:20–21).[78] "Religion that is pure and undefiled before God" is defined in part as caring for "orphans and widows in their distress" (James 1:27). Yet Christian love was not to be extended merely to fellow Christians but to neighbors and even enemies. When Jesus was asked, "And who is my neighbor?" he responded by relating the parable of the good Samaritan (Lk. 10:25–37). When a Jewish man lay on the road from Jerusalem to Jericho, having been attacked by highwaymen and needing medical attention, a Levite and a priest each passed him by and refused to give him assistance, thereby disgracing their own moral standards, which required them to care for their own. While Jews tended to look down on Samaritans, it was a Samaritan who showed himself to be a neighbor in the sense that the wounded man's own countrymen had failed to be: he had compassion on him *(esplanchnisthē)* and he gave him medical aid. Glanville Downey maintains that the concept of *agape* that underlies Jesus's parable marked a radical innovation if we compare it with classical responses that would have been given to the question that was posed by Jesus. In place of a Stoic doctrine of human brotherhood or a definition of the nature of humanity, it grounded philanthropy in a theological conception that saw human love as reflecting divine love.[79] But it also went beyond Jewish concepts of charity, which was directed inward to one's own community. The novelty of Jesus's teaching was that beneficence extends beyond one's own community. His command was, "Go and do likewise" (v. 37). In several passages in the Gospels Jesus enunciates the pattern of personal charity that was to be incumbent on his followers. "[F]or I was hungry and you gave me food, I was thirsty and you gave me something to drink, I was a stranger and you welcomed me, I was naked and you gave me clothing, I was sick, and you took care of me [*epeskepsasthe*], I was in prison and you visited me. . . . [J]ust as you did not do it to one of the least of these, you did not do it to me" (Mt. 25:35–36, 45). The verb *epeskepsasthe* (from *episkopein*), used in this passage for taking care of the sick, is sometimes employed in late classical Greek to describe a physician's visiting a patient.[80]

It is not difficult to see the gap that existed between the classical concept of *philanthropia* and the Christian idea of *agape* as an ethical dynamic. Nor is it surprising that philanthropy has been called a peculiarly Christian product.[81] While Christian philanthropy had its roots in Judaism, the concept of *agape* led to a broadening and deepening of the Jewish impulse, especially in its not being limited to the believing community. By the end of the second century, *philanthropia* began to appear frequently in the Christian vocabulary, perhaps because it was a word (unlike *agape*) that pagans could readily understand.[82] It is often used by the church fathers to describe God's love for humanity as shown in the Incarnation. By the fourth century it came to be used as a synonym for *agape* in the liturgies of the Greek church.[83]

A second consequence of the doctrine of the *imago Dei* was that it provided the basis for the belief that every human life has absolute intrinsic value as a bearer of God's image and as an eternal soul for whose redemption Christ died. This belief led to a stern and uncompromising condemnation of pagan morality in all its aspects. Christians viewed its tolerance of the elimination of unwanted human life and of the cruelty shown to those whom society had condemned or abandoned as an indication that Roman society was incurably wicked. They attacked abortion, infanticide, the gladiatorial games, and suicide in the strongest possible terms.[84] Early Christians showed special concern for the protection of unborn and newborn life. Abortion, though occasionally condemned in classical antiquity,[85] was widely practiced, and the fetus, being regarded as part of its mother, enjoyed no legal protection or absolute value until the third century, when abortion was penalized by a rescript issued under the emperors Septimius Severus and Caracalla between 198 and 211.[86] As early as the second century we find abortion condemned in Christian writings for violating God's handiwork. In the Didache the aborted fetus is called a "molded image [*plasma*] [of God]" (2.2). In the second-century Apocalypse of Peter abortion is said to corrupt "the work of God who created" it. This theme is reiterated in the numerous examples of denunciation of abortion that are found in the church fathers.[87] The difference in Christian and pagan attitudes toward abortion reflected a difference in how the fetus was perceived.[88] Pagans considered the victims insignificant; Seneca thought that to drown a newborn was an act of reason, not of anger.[89] To Christians, however, the fetus was not only human but an eternal soul. Abortion was regarded by some as worse than murder. Tertullian explicitly calls abortion homicide: "For us, indeed, as homicide is forbidden, it is not lawful to destroy what is conceived in the womb while the blood is still being formed into a man. To prevent being born is to accelerate homicide, nor does it make a difference whether you snatch away a soul which is born or destroy one being born. He who is man-to-be is man, as all fruit is now in the seed."[90]

The exposure (abandonment) of newborn children was also condemned in early Christian writings.[91] Whether or not it was forbidden by law under the empire (and this is disputed), it was not punished, and it was widely practiced and viewed with general indifference.[92] Exposure was attacked by Christians,[93] who viewed it as a crime. Christians also emphatically condemned suicide, which had been idealized in classical antiquity as a noble means of death.[94] Believing that they ought to endure suffering with the help of God's grace rather than seek to put an end to their lives, Christians regarded suicide as self-murder. Augustine discusses the matter at length,[95] for the most part summarizing the views of early Christian writers. The only serious debate over the propriety of suicide involved cases in which a woman's chastity was in danger, on which Augustine differed from earlier writers. His condemnation of suicide (on the ground that it is homicide and precludes the possibility of repentance) proved to be authoritative in the early church.[96]

A third consequence of the doctrine of the *imago Dei* was in providing early Christians with a new perception of the body, and indeed of the human personality.[97] Late pagan proponents of asceticism went beyond the earlier Greek concept of *askēsis*, or training of the body. They expressed no admiration or concern for the body; on the contrary, they were ashamed of it. They looked forward to the day when at death the soul would free itself from matter, which they regarded as evil.[98]

The Greek dualism of the body-soul dichotomy was taken over by Gnostics, who wished (like pagan ascetics) to free themselves from their own bodies.[99] But orthodox Christians did not adopt Gnostic or Manichean dualism.[100] Christians generally viewed asceticism as a means of strengthening the body in the struggle against demonic forces, not of mortifying it.[101] It was just at this point that Christian ascetics differentiated themselves from the familiar type of the pagan ascetic.[102] The dichotomy between the material body and the spiritual soul provided the philosophical basis for the pagan rejection of the Christian doctrine of the Incarnation: How could a spiritual being (God) take on corruptible flesh?[103] For the Christian the Incarnation provided the ground for salvation: the eternal God had become man in order to save the human race through his death and resurrection.[104] By his death the human race gained redemption, by his resurrection eternal life.

A new perception of the body led to the formulation of a novel concept of personhood that provided the theological basis for integrating body and soul in a manner that was unknown to either Platonists or Stoics.[105] Christ served as the exemplar of this integrated personality, combining within himself the two natures of God and human. The Christian conception of Jesus as perfect man contributed to raising the body to a status that it had never enjoyed in paganism.[106] Docetism (the view that Jesus's humanity was apparent rather than real) was an attempt by Gnostics and others to escape the idea, which was repugnant from a traditional pagan or a Gnostic point of view, of a material body being absorbed into the spiritual Godhead, as orthodox theologians posited. In their rejection of Docetism orthodox Christians insisted that the body was not evil; if the Son of God had assumed a true body ("truly God and truly man"), then it must be, like all the material cosmos, good. In place of the dualism of Greek philosophy Christian doctrine posited a new divide: between the old humanity, in which both body and soul were tinctured by original sin, and redeemed humanity, in which both body and soul were cleansed of sin and the divine image that had been implanted in them was restored.[107] It was to save the *body* that Christ took on flesh in the Incarnation.[108] Not only the soul, which in traditional pagan thought was eternal, but the composite of body and soul, which constituted man, was to be resurrected,[109] an idea that was as repellant to pagans as the doctrine of the Incarnation.[110] Classical philosophers envisioned a continuum between the soul and God and a divine immanence that pervaded the cosmos.[111] In marked contrast, Augustine saw a chasm between the soul and God that could be bridged only by the incarnate Christ, who had at a particular moment entered the temporal dimension, an idea that was itself disturbing to pagans.[112] The divine compassion is mirrored in a human compassion for others, which becomes the basis of ethics and a means of reclaiming the *imago Dei* in humans.[113] Hence the Christian understanding of personhood became the foundation for a new series of relationships in which the Christian community (*civitas Dei*) would come to supplant the classical polis (*civitas terrena*) as the focus of human activity.[114] This community, the larger metaphorical "body of Christ," consisted of all believers—Jew and Greek, slave and free, male and female (Gal. 3:28)—who formed a unified body in Christ and as such were members of one another (1 Cor. 12:5).[115] Here indeed was a unique concept of the human personality—a psychosomatic unity, a composite of body and soul—which created new boundaries that transcended traditional political and social divisions.[116]

A fourth consequence was that the doctrine of the *imago Dei* led to a redefinition of the poor. The human body in all its parts shared in the divine image.[117] This was true of the bodies not merely of Christians but of all people. It was true particularly of the poor, who acquired a new definition in Christian thought: those who had true worth because they bore the face (*prosōpon*) of Christ.[118] According to Susan Holman, the theology that lay behind the new prominence that Christians accorded the poor was specifically Nicene rather than Arian.[119] The Cappadocian fathers constructed an identity of the poor based on the belief that Jesus was the incarnate God, a belief that imparted a redemptive nature to early Christian relief efforts.[120] As human beings (*anthrōpōn anthrōpoi*) they shared, in the words of Gregory of Nyssa,[121] a common nature (*koinēn phusin*). Even the diseased body of a leper had importance. Like Lazarus, to whom lepers were frequently compared by the Cappadocians, they are sanctified because they bear the image of their Savior. No longer repulsive, they bring holiness and healing from spiritual diseases to those who touch them in order to assist them.[122] "By taking the lepers' flesh in hand, those who minister to them participate in the divine immanence of creation that proceeds from the incarnate Son's essential sharing in both deity and cosmos."[123] The new image of the poor did not reflect a Christian romanticizing of their condition. But it did constitute a challenge to the rich and powerful, who had traditionally claimed to merit a special relation with the gods in their role as patrons of the community.[124] Sermonic literature depicts Jesus as having chosen in his Incarnation to identify himself with the poor rather than with the rich, since the former could boast of no advantages that gave them a claim to his favor.[125]

But the image that lay behind the doctrine of the Incarnation went beyond a mere class division between the rich and the poor. Christ had united in his own body a wider chasm, which separated the high state of God and the abject poverty of the human condition.[126] The mystery of the Incarnation united heaven and earth and formed the basis of a "new language of solidarity," the solidarity of members of "the body of Christ." Participation in the Eucharist allowed every Christian to share in the Savior's divine flesh and provided a means for incorporating humanity into the larger mystery of his spiritual body, the church.[127] Just as God demonstrated in the Incarnation his solidarity with those who suffer, so the members of his "body" must demonstrate their solidarity with the suffering poor.[128] The social implications of theological formulations were very much a part of the Christological disputes that arose regarding how best to express the relation between the human and the divine in the person of Christ. The Monophysites claimed that Christ possessed a single nature that merged the divine and the human, while the Chalcedonians held that he possessed two separate natures, divine and human. The Monophysite party arose after the Council of Chalcedon in 451, which declared the doctrine of the two natures the orthodox one. In the debates each party saw the other's formulation as endangering the importance of God's compassion for the wretched poor, whose flesh Christ shared and for whom his spiritual body on earth ought to care by acts of mercy.[129] The language inherent in these formulations, and in the sermons that drew their inspiration from them, reflected an underlying theology that saw the Incarnation as the basis for compassionate care of those in need.[130]

The Christian Physician

While all Christians who wished to do so could endeavor to pursue a life of *imitatio Christi*, the Christian physician who desired to serve God had a well-developed ideal to emulate. Early Christian authors adopted as compatible with the New Testament's emphasis on beneficence the classical pagan tradition that employed, in simile and metaphor, the idea of the physician as a compassionate, selfless, and philanthropic healer of ills and other forms of distress.[131] Owing to the high associations that had become attached to the word in classical literature, the physician as an ideal provided a commonplace for early Christian homiletics, with the church fathers often drawing on the practice of medicine for spiritual analogies. Jesus became "the true physician" (*verus medicus, verus archiater*) and "the only physician" (*solus medicus*), and was described as "himself both the physician and the medication" (*ipse et medicus et medicamentum*).[132] The qualities, long associated with Hippocrates and Asclepius, of the ideal physician who unselfishly succors the ill came to be attributed by Christians to him. A highly idealized Hippocrates was adopted by some Christian authors as an exemplar of virtue. Indeed, Jesus is himself called "a spiritual Hippocrates" (*quasi spiritualis Hippocrates*).[133] The symbolism is no less potent for describing healing of the soul rather than of the body. Rich in its association with compassion and concern, medicine became one of the most appealing and widely used analogies for the Christian cure of souls.

Vivian Nutton argues that Christians considered the profession a pagan one and the speculative philosophy of Galen and Hippocrates dangerous to Christians. "An ambiguity towards pagan medicine at a popular level," he writes, "contributes to a certain suspicion of doctors at a higher level."[134] Yet some Christians had little trouble with Galen, whom they came to appreciate for his prolific output of medical and philosophical works. Robert Grant thinks it likely that Origen had read several of Galen's medical and philosophical treatises,[135] and Jerome seems to have done so as well.[136] Moreover, during the pontificate of Victor as bishop of Rome (c. 189–c. 198) a group of Christians led by Theodotus of Byzantium, who advocated an adoptionist or Monarchian Christology, were influenced by Galen in evidently attempting to present Christianity in philosophical terms that would appeal to pagans.[137] While it is true that Christians initially condemned pagan philosophy, they came in the second century to borrow extensively, if selectively, from it. Galen was the first pagan writer to treat Christianity with respect as a philosophy rather than, like most educated Romans, as a superstitious sect. He admired Christians for their contempt of death, their sexual purity, self-control in regard to food and drink, and their pursuit of justice, in all of which he regarded them as not inferior to pagan philosophers. He criticized Christians, however, along with Jews, for their refusal to base their doctrines on reason rather than on faith and revealed authority. Some Christians (e.g., Nemesius, Isidore of Pelusium) criticized Galen for his philosophical naturalism, which led him to deny the immortality of the soul. But Christians were not alone in their criticism, in which the Neoplatonic philosopher Proclus joined. Beginning in the fourth century several Christian writers demonstrate the influence of Galen's medical and philosophical views.[138]

We have few references to Christian physicians in literary sources of the early centuries of Christianity.[139] Apart from Luke, the writer of the Third Gospel and Acts,[140] we have no specific literary reference to any Christian physician before the late second century, when we hear of a Phrygian physician, Alexander, who was martyred at Lugdunum (Lyons) in Gaul around 179, during the reign of Marcus Aurelius.[141] Eusebius mentions a second physician who was martyred in the persecution of Diocletian.[142] While the literary evidence is spotty, epigraphic and papyrological sources supplement literary sources as evidence of early Christian physicians. Christian Schulze has compiled a census of every physician in classical antiquity who is identified as a Christian.[143] Some 150 physicians can be identified with reasonable certainty from the Roman imperial period.[144] While most can be dated to late antiquity, his census of 194 persons includes some physicians who lived as late as the early eighth century. Four are from the first century, 5 from the second, 19 from the third, and 32 from the fourth, together with an additional 35 who can be dated to the fourth or fifth century, although there are some in each category whose identity is not certain. In addition to enumerating Christians who practiced medicine, Schulze provides valuable comparative data. He lists 59 trades and professions that early Christians are known from epigraphic evidence to have entered[145] and observes that no other professional group comes close to the number of physicians that are mentioned in the Latin inscriptions alone. In fact, the number of Christians in other trades (Schulze uses bakers as an example) is relatively small. The sample that he uses for comparison includes about 90 doctors and 27 bakers. Far from being rejected as a profession, Schulze asserts, medicine proved to be an especially attractive one to Christians.[146] A precise enumeration century by century permits us to correlate very roughly the increase in the number of Christian physicians with the increase in the Christian population. Rodney Stark concludes that there were about 7,530 Christians by the end of the first century A.D. (Schulze identifies 2 definite and 2 uncertain physicians from the first century); 217,795 by the end of the second (Schulze identifies 4 who are definite and 1 uncertain physician from the second century); and 6.3 million by the end of the third (Schulze identifies 15 definite and 4 uncertain physicians from the third century). [147] While different assumptions will produce different population figures, those of Stark give us a fair idea of the *rate of growth* of Christianity. We should expect the number of Christians who can be identified as physicians to have been relatively small as long as Christians were persecuted (from A.D. 64 to 313). But after the legalization of Christianity by Constantine in 313, the number of Christians grew rapidly,[148] as did the number of physicians who can be identified as Christian (Schulze identifies 32 definite and 4 uncertain physicians from the fourth century and an additional 23 definite and 12 uncertain physicians from either the fourth or the fifth centuries). [149] It is unlikely that there existed—as has been alleged—a widespread Christian suspicion that the medical profession was inherently pagan. To be a pagan physician was to be a servant of Asclepius, but not all physicians viewed Asclepius's cult with equal favor, and loyalty to his cult did not pose a barrier to Christian conversion.[150] Nor did Asclepius's patronage render medicine pagan. Because of its naturalistic approach to healing, it was value-neutral and could be practiced by those of any or no religious beliefs. Medicine was no different from any other craft or guild in having its own patron deity. The widespread appeal of the "Christus medicus" motif made it possible for pagan physicians who became Christians to secure an equally compassionate patron to replace Asclepius. Nor should we view the close link between medicine and philosophy as an insuperable impediment. The familiarity that many of the fathers exhibit with Greek medical writings demonstrates that in spite of Theodotus's heresy Christians did not ordinarily regard medical theory as theologically harmful, and it did not hinder Christians from becoming physicians.[151] In fact, however, few physicians—and they will have been mostly status-conscious upper-class physicians—became philosophers in the sense that Galen recommended.[152]

When hospitals were established in the late fourth century, they sometimes included physicians. Some Christians, in fact, began the practice of medicine as a vehicle for Christian charity. A new model of the Christian physician was set forth in the highest terms by Christian writers, which drew on the examples of both Hippocrates, the ideal physician, and Christ, the healer of spiritual ills. In their care of the destitute and the poor, physicians evinced the charitable spirit of Christianity, as Origen wrote that Jesus had earlier attempted to imitate "the method of a philanthropic physician who seeks the sick so that he may bring relief to them and strengthen them."[153] The physician, he writes, manifests a Christ-like compassion in his care for the commoner, the destitute, and the poor. Augustine, too, writes of the ideal physician who is motivated by charity and hence seeks no remuneration for his services, treating the most desperate cases among the poor with no thought of a reward.[154]

It is difficult to determine with any degree of accuracy what influence these ideals had on the actual practice of medicine.[155] Our evidence is largely anecdotal, though there is enough to suggest that they enjoyed a growing influence after the legalization of Christianity in 313. The extent of physicians' conformity to the Christian ideal in medicine is likely to have been proportional to their Christian conviction and commitment. Moreover, even for Christian physicians the role of the physician was defined not by Christian ideals but by Hippocratic precepts that had long been enshrined in the Hippocratic Corpus. With the exception of issues like abortion, exposure, and assisted suicide, the medical ethics of Christian physicians are not likely to have been defined very differently than were those of their pagan colleagues, except perhaps for a greater willingness to help the poor.[156] Some physicians, however, attempted to combine commitments to faith with those to the traditional expectations of the medical art in such a way as to carry out the ideals of Christian philanthropy. Thus Augustine points to his friend, the physician Gennadius, whom he describes as a man "of devout mind, kind and generous heart, and untiring compassion, as shown by his care of the poor."[157] Zenobius was a fourth-century priest and physician who is lauded by his biographer for serving his poor patients without remuneration and even helping them financially when necessary.[158] Augustine believed that the physician should always be concerned for the cure of his patient,[159] for if the physician were merely concerned about the practice of his art, medicine would be cruelty.[160] But descriptions like these are hardly novel. In fact, they can be paralleled in honorary inscriptions of pagan physicians who aided their communities in difficult circumstances in both the Hellenistic period and Roman imperial times.[161] Even more specifically Christian are the elements of caring for the poor and the destitute, the conjunction of the care of the body with the care of the soul, and compassion.

Some human qualities seem so instinctive that we presume, perhaps too readily, that they spring naturally from the human heart. Compassion is such a quality. The English word is derived from the Latin verb *compati*, which has reference to suffering or having sympathy with someone.[162] The word *patient* is derived from the same root. The term *compassion* (Gk., *eusplanchnia*; L., *misericordia, eleemosyna*) calls to mind such cognate ideas as pity, mercy, sympathy, and beneficence.[163] It includes being "a lover of the destitute" (*philoptōchos*), which is a specifically Christian concept that is rarely found in classical Greek usage but appears in the fourth century.[164] It denotes an intuitive identification with the pain and suffering of another person.[165] The recognition and spread of compassion as a definable Christian virtue that could be applied to the practice of medicine are likely to have required a leavening process, lasting for several generations, by which Christian theology and patterns of thought came to permeate the larger society and inform its values. These patterns and the theology that lay behind them did not supplant the older Hippocratic tradition of the ideal physician, which retained a strong influence. Rather, they supplemented it by enjoining specifically Christian ideals in the practice of medicine. In the fourth century we find increasing mention of clergy (priests and monks) whose spiritual and medical interests blended into a common concern for the spiritually and physically ill. An example is Hypatios, a monk and a physician in the late fifth and early sixth centuries who, according to his biographer, treated patients afflicted with various sores who came to him because, being poor, they had been refused treatment by other physicians.[166] Basil speaks of Eustathius, a physician who combined a spiritual ministry with his medical practice. He writes (c. 375), "And your profession is the supply vein of health. But, in your case especially, the science is ambidextrous, and you set for yourself higher standards of humanity, not limiting the benefit of your profession to bodily ills, but also contriving the correction of spiritual ills."[167] It is not for combining secular and religious means of treating physical ailments that Basil praises Eustathius but rather for the effect of his Christian ideals on his "standards of humanity" and his concerning himself not only with treating the physical ills of his patients but also with ministering to their spiritual ills. From this and other sources (which become abundant in the fourth century) there emerges a picture of the ideal Christian physician who combines the medical art with spiritual commitment. But the sources do not suggest that physicians attempted to adopt any form of miraculous or ritual healing, thereby discarding secular medicine for supernatural means. There is relatively little evidence for the sixth and seventh centuries, but more for the eighth and ninth, for the practice of medical charity by the clergy, especially by monks.[168] Monasteries became the refuge of the persecuted, the poor, and the sick. In many instances such care as could be provided was given by monks who made no claim to extensive medical knowledge and would hardly, even by the standards of the time, be considered physicians. But examples survive as well of priests who were regarded as *medici.*

We must be careful here to recognize that clerical or monastic physicians were first and foremost clergy, for whom the practice of medicine was an extension of their monastic role, an act of Christian charity performed for the glory of God and the love of the human race. But it was not sacerdotal medicine of the sort practiced by priests in societies in which the supernatural etiology of disease dictated a reliance upon supernatural means of treatment. The medicine practiced by monastic and clerical physicians in late antiquity and the early Middle Ages, although by modern standards riddled with simplistic, erroneous, and sometimes superstitious explanations and procedures, was not primarily magical or religious. The religious functions of these physicians would, at the most, be complementary to their medical efforts and were often directed as much toward the patients' spiritual as to their physical ills.

Hippocratic and Christian Ideals of Medical Practice

"The Greek doctor," writes W. H. S. Jones, "was not compelled to act properly; he was merely trained to consider right behavior as 'Good Form' (εὐσχημοσύνη). Such a sanction allows rules to be general and vague. . . . The Greek doctor . . . was an artist first and a man afterwards."[169] By "artist" Jones means a craftsman, whose practice of medicine reflected a devotion to his art, not to an ethical standard. In contrast, the Christian physician (in principle, at any rate) considered his first obligation to be to God, while he viewed his patient as one who bore God's image. This understanding allowed him less latitude in his approach to ethical issues than the pagan physician enjoyed. It is for this reason that the so-called Hippocratic Oath appealed to Christians.

Precisely when the oath was written is unknown.[170] Although it is first mentioned by Scribonius Largus in the first century of our era, it may date from as early as the fourth century B.C.[171] At first glance it seems to offer a very different approach than that of the deontological treatises of the Hippocratic Corpus. Its religious tenor and some of its injunctions (e.g., prohibition of abortion, euthanasia, and perhaps surgery) indicate that it originated among a restricted group of physicians. Ludwig Edelstein suggested the Pythagoreans, who belonged to a philosophical sect that emphasized moral purity, asceticism, and piety and who lay outside the mainstream of Greek medicine.[172] Those who took the oath swore by Apollo, Asclepius, and other gods and goddesses of healing to guard their life and art "in purity and holiness."[173] The oath was regarded by some pagan medical writers during the Roman imperial period as setting forth an ideal standard of professional behavior, but at no time was it used in the classical world to regulate the practice of more than a minority of physicians.[174] There was much in the oath, however, that appealed to Christians: its religious tenor, its prohibition of abortion, and its standard of sexual purity.[175] As a result some Christian physicians appropriated the oath's precepts and infused them with new meaning.[176] Since their obligation had to be stronger and more binding because it was informed by their service to God, the oath became more than a counsel of perfection or an ideal to be striven for. The Christian physician's conscience needed to be bound by a formal affirmation. Hence while Hippocratic medical etiquette was taken over mutatis mutandis by Christians (as it was later by Jewish and Muslim physicians), the oath seems to have acquired a role that was lacking in its pre-Christian use.[177]

In the sixth century Cassidorus (c. 487–583) penned two documents describing the duties of physicians that testify both to the persistence of Hippocratic medical ethics in late antiquity and to the introduction of a specifically Christian emphasis on compassion in medicine. The first of these was composed when he was in the service of the Ostrogothic king Theodoric (493–526) in Italy. This document[178] reinstated the office of *comes archiatrorum*, who appears to have been both the president of the college of civic physicians in Rome and personal physician to the king and the royal household. The text begins with an encomium on the usefulness of the art of medicine, which is labeled "glorious" because it drives out diseases and restores health. Cassiodorus lauds the nearly uncanny prognostic skill of experienced physicians, praises the art for being a learned discipline, and admonishes physicians to trust its science rather than their own experience. He also chastises physicians for their bedside bickering, urges them to work together in assisting each other harmoniously, and reminds them that at the beginning of their career they swore an oath to hate iniquity and to love purity. This document breathes the spirit of Hippocratic medical ethics and could just as easily have been composed centuries earlier by a pagan writer. Little in it is distinctly Christian.

After retiring from the court of Theodoric, Cassiodorus founded a monastery at Vivarium in Bruttium. In his *Introduction to Divine and Human Readings* he writes to those of his monks who were also physicians:

> I salute you . . . who are sad at the sufferings of others, sorrowful for those who are in danger, grieved at the pain of those who are received, and always distressed with personal sorrow at the misfortunes of others, so that, as experience of your art teaches, you help the sick with genuine zeal; you will receive your reward from him by whom eternal rewards may be paid for temporal acts. Learn, therefore, the properties of herbs and perform the compounding of drugs punctiliously; and do not trust health to human counsels. For although the art of medicine is found to be established by the Lord, he who without doubt grants life to men makes them sound. For it is written: "And whatsoever you do in word or deed, do all in the name of the Lord Jesus, giving thanks to God and the Father by Him."[179]

It is instructive to compare this exhortation with that which he had directed to the *comes archiatrorum* and, by extension, to physicians in the public medical service. In both documents Cassiodorus praises the art of medicine. But while in the first he merely urges the secular physicians, in classical fashion, to place their confidence in their art, in the second he encourages the monks to place their hope in the Lord rather than in the art of medicine. Cassiodorus's guidance to secular physicians largely reproduces the traditional Hippocratic virtues expected of physicians. Although he writes that secular physicians are to be dedicated to their art and mindful of the oath that they swore, he places a minor emphasis on the calling, motivation, and qualities of the secular physician. Presumably he does so because he is writing what is essentially a secular document. In contrast he urges the monk-physicians to be motivated by compassion to "perform [what he terms] the functions of blessed piety" for a reward that will be bestowed by the Lord. In the medical-ethical literature of the early Middle Ages, these religious and philanthropic ideas of monastic medicine were merged with the earlier secular tradition of Hippocratic medical ethics.[180] Both came to form important strands in the tradition of Western medical ethics. With the introduction of the Christian emphasis on compassion as an essential motive, one can speak of something new in medical ethics—an element that cannot be said to have represented an ideal in pre-Christian medicine. Jesus's parable of the Good Samaritan became the model of Christian *agape*. Compassion—not merely duty to the art—became the motivating ideal of the Christian physician. Nevertheless, the fact that Cassiodorus encourages the spirit of compassion in monks who were physicians while omitting to mention it in writing to the civic physicians (who did not necessarily practice medicine from religious motives) suggests that it was too much a matter of the heart to be an enforceable criterion for admission to the practice of medicine.

Conclusion

Graeco-Roman society recognized philanthropy as a motive for the practice of medicine, but it was never essential to the ideal of the classical physician. The meaning of the concept changed over time: in *Precepts* it is kindliness; for Galen, being a philosopher. Galen, in surveying the several motives that might attract individuals to engage in the medical art, recognized a variety of incentives, of which philanthropy was one, as were desire for money, honor, and immunities from taxation. But only competence was essential. In excluding pity as a basis for personal assistance, classical philanthropy differed markedly from Christian charity in both motive and practice.[181] The Stoic conception of *apatheia* (insensibility to suffering), moreover, encouraged an attitude of quietism that was content to accept the world as it was rather than to try to change it. While it would be presumptive to doubt that many pagan physicians exercised compassion in medical treatment, there existed in the classical world no external impetus, no elevated ideal, no specific virtue, of compassion. With rare exceptions (e.g., Scribonius Largus), ancient philosophical or medical writers did not expect the virtuous physician to be humanitarian or philanthropic in the practice of medicine. That expectation had to await the coming of Christianity, which substituted the idealization of very different virtues for those that had long dominated the classical world.[182] In the medico-ethical literature of the early Middle Ages the new religious and philanthropic ideals of monastic medicine were merged with the older secular tradition of Hippocratic medical ethics and etiquette. In words that Susan Holman uses to describe a different appropriation by Christians of Greek ideals, "The act of adaptation functioned as both symptom and cause: symptom of the general atmosphere of intersecting ideologies, and cause of the newly ordered identities that evolved."[183]

CHAPTER SIX

Health Care in the Early Church

The development of the care of the sick in early Christianity has sometimes been viewed as having occurred in two stages. In the pre-Constantinian church, charitable activity, including the care of the sick, depended largely on the ministrations of nonmedical clergy and laity. After the legalization of Christianity in A.D. 313 and the influx of state funds that came to be directed to its support, the creation of permanent medical institutions marked the decline of a congregation-centered approach in favor of organized institutional efforts on behalf of those requiring medical treatment. I shall argue in this chapter that this view does not do justice to the evidence, which indicates that a more highly developed pattern of charity existed in the early church than is sometimes recognized. In particular, I shall suggest that by the third century the rapid growth of Christianity in the cities of the Roman Empire led to the parochial organization of benevolent work on a large scale.[1] The plague of Cyprian, which beset the empire in the mid-third century, greatly extended its philanthropic role, marking a considerable advance over the organization of medical charity that preceded it and preparing the way for permanent nonparochial medical institutions, especially hospitals, in the fourth century. But the creation of the hospital did not end the role of the urban churches in administering medical charity, which continued for several centuries.

The Organization of Medical Care

From the very beginning Christianity displayed a marked philanthropic imperative that manifested itself in both personal and corporate concern for those in physical need. In contrast to the classical world, which had no religious impulse for charity that took the form of personal concern for those in distress, Christianity regarded charity as motivated by *agape*, a self-giving love of one's fellow human beings that reflected the incarnational and redemptive love of God in Jesus Christ. At the same time that ordinary Christians were encouraged privately to visit the sick and aid the poor, the early church established some forms of organized assistance.[2] The administrative structure of the local church (*ecclesia*) was simple but well suited to the supervision of charitable activities that relied largely on voluntary activity. Each church had a two-tiered ministry composed of presbyters (priests) and deacons (see Acts 6:1–6), who directed the corporate ministry of the congregation. Deacons, whose main concern was the relief of physical want and suffering, had a special duty to visit the ill and report them to the presbyters: "They are to be doers of good works, exercising a general supervision day and night, neither scorning the poor nor respecting the person of the rich; they must ascertain who are in distress and not exclude them from a share in church funds, compelling also the well-to-do to put money aside for good works."[3] Collections of alms were received every Sunday for those who were sick or in want.[4] They were administered by presbyters and distributed by deacons. Widows who did not need assistance formed a separate class that was later replaced by the office of deaconess. These widows and deaconesses were expected to help the poor, especially women, who were sick.[5] Although their numbers and resources might be small, Christians were equipped, even in the most adverse circumstances, to undertake considerable charitable activity on behalf of those who were ill. Owing to a combination of inner motivation, self-discipline, and effective leadership, the local congregation created in the first two centuries of its existence an organization, unique in the classical world, that effectively and systematically cared for its sick.

In the third century the rapid growth of the church, particularly in the large cities of the Roman Empire, led to the organization of benevolent work on a large scale. Roman cities were crowded, often unsanitary, and, for large numbers of city dwellers, lonely. There existed groups, like guilds and burial societies, that maintained a degree of fellowship and mutual support, but there were many urban dwellers who were outside any family or social network of support. As the number of those who benefited from the church's charitable activity increased, there came to be too few clergy to deal with the demands made on them. Churches were reluctant to appoint more than seven deacons, the number thought to have been chosen by the apostles (see Acts 6:1–6, if indeed the passage describes deacons). Hence congregations began to create minor clerical orders to assist them, such as subdeacons and acolytes. In a letter that is preserved by Eusebius, written in 251 by Cornelius, bishop of

Rome, to Fabius, bishop of Antioch, we learn that the church in Rome supported 46 presbyters, 7 deacons, 7 subdeacons, and 42 acolytes, as well as 52 exorcists, readers, and doorkeepers—altogether a staff of considerable size.[6] Apparently the church in Rome had divided the city into seven districts, each under a deacon, who was assisted by a subdeacon and six acolytes. They cared for 1,500 widows and distressed persons who were supported by the church.[7] Adolf Harnack estimated that the Roman church spent each year from 500,000 to 1 million sesterces on the maintenance of those in need.[8] We know that as early as the second century the church at Rome had large sums at its disposal. When Marcion came to Rome from Pontus around 139, he made a donation of 200,000 sesterces to the church, which was returned several years later when he was excommunicated.[9] The fact that the church was able to return such a large sum of money furnishes a good indication of the resources at its disposal. Other churches were able to raise large subscriptions at short notice. In the third century Cyprian relates that the Carthaginian churches contributed 100,000 sesterces to the Numidian churches in order to redeem local citizens who were being held for ransom.[10] A century later John Chrysostom writes that the great church in Antioch supported 3,000 widows and virgins along with other sick and poor persons and travelers.[11]

All this—the establishment of minor orders to assist presbyters and deacons, the creation of sizable staffs of clergy in large churches, the regular support of considerable numbers of the poor and sick, and the expenditure of large sums of money—suggests that the churches devoted a good deal of attention to corporate philanthropic activity. The maintenance of the sick was viewed by the pre-Constantinian churches as a part of their charitable ministry. As that ministry grew, so apparently did the number of sick who were supported. Presumably much of the care was directed toward relieving individual suffering rather than rendering prophylactic or therapeutic treatment, and it is likely that the assistance given was in many cases rudimentary and palliative. The church's care of the sick relied primarily on the clerical orders, which were composed of men chosen for their spiritual rather than medical qualifications. If they possessed the latter, it would have been merely incidental.

The Plague of Cyprian

In A.D. 250, a plague spread throughout the Roman Empire that called for a much more extensive effort than churches had previously put forth on behalf of the sick. Commonly called the plague of Cyprian, it is said to have originated in Ethiopia and to have spread rapidly through Egypt to North Africa and thence to Italy and the West as far as Scotland, where it reached epidemic proportions.[12] It recurred at intervals in the same district, with brief remissions that were followed by additional severe attacks.[13] It lasted for fifteen or twenty years and carried off large numbers of the population of the Roman Empire. According to Zosimus, the mortality rate was higher than in any previous epidemic.[14] In some places the number of those who died outnumbered survivors. In Rome 5,000 people are said to have succumbed in a day.[15] No real understanding of public hygiene existed in antiquity.[16] Health regulations existed chiefly for aesthetic, not sanitary, purposes (to rid cities, for example, of the foul odor of sewage). Sewage was sometimes stored in close proximity to wells, increasing the likelihood of diarrhea and dysentery, which are frequently mentioned by medical writers.[17] Other diseases that resulted from poor environmental conditions included cholera, gastroenteritis, infectious hepatitis, leptospirosis, and typhoid.[18] Except for making supplications to the gods, the civil authorities did little to alleviate the situation.[19] Responsibility for health was regarded as a private, not a public, concern. In spite of well-known epidemic diseases in the ancient world (e.g., the plague of Athens [430–29 B.C. with recurrences], the Antonine plague [A.D. 166–72 with recurrences], the plague of Cyprian [A.D. 250–c. 270], and the plague of Justinian [A.D. 541–749]), most outbreaks of infectious disease were left to individuals to deal with on a self-help basis.[20] Emergency measures were rarely taken by municipal officials—hence the frequently described scenes in classical literature from Thucydides to Procopius of corpses lying unburied in the streets during times of plague.[21] Thucydides, in a well-known passage that became a model for later writers, described the plague of Athens in 430 B.C. in a city overcrowded with citizens of outlying villages who had taken refuge inside the walls during the Spartan invasion of Attica:

> Terrible, too, was the sight of people dying like sheep through having caught the disease as a result of nursing others. This indeed caused more deaths than anything else. For when people were afraid to visit the sick, then they died with no one to look after them; indeed, there were many houses in which all the inhabitants perished through lack of any attention. . . . The bodies of the dying were heaped one on top of the other, and half-dead creatures could be seen staggering about in the streets or flocking around the fountains in their desire for water. The temples in which they took up their quarters were full of the dead bodies of people who had died inside them.[22]

In the classical world there was little recognition of social responsibilities on the part of the individual.[23] Before the advent of Christianity, moreover, there was no concept of the responsibility of public officials to prevent disease or to treat those who suffered from it. Alex Scobie speaks of "a cynical acceptance of the state's indifference to the lot of the urban poor."[24] In part, this can be explained by the belief in pollution (*miasma*) and purification (*katharsis*). The general acceptance of calamities as the retribution of the gods that indicated their displeasure was deeply rooted in Greek and Roman religion and remained a part of paganism until the end of antiquity. Plague was attributed to the gods, who punished men for having violated a taboo or incurred divine displeasure by bringing pollution on a city, whether intentionally or unintentionally—but not for moral offenses, since the gods imposed no ethical requirements. Only public sacrifice or purification could satisfy the anger of the gods. It remained the responsibility of magistrates as religious representatives of the city to determine the reason for a plague and to supplicate the gods to bring about its end. Traditional attitudes of pessimism and quietism—the feeling that little could be done on a public level to end widespread disease or to care for the ill—underlay the inactivity of public officials and their failure to undertake strenuous measures. The fact that many outbreaks of infectious disease were local and often associated with famine or siege meant that greater problems absorbed their attention.[25] Eusebius vividly describes an epidemic that broke out in A.D. 312–13, during the reign of Maximin Daia, following a drought-induced famine:

> It was the winter season, and usual rains and showers were withholding their normal downpour, when without warning famine struck, followed by pestilence and an outbreak of a different disease—a malignant pustule, which because of its fiery appearance was known as a carbuncle. This spread over the entire body, causing great danger to the sufferers; but the eyes were the chief target for attack, and hundreds of men, women, and children lost their sight through it. . . . In the Armenian war the emperor was worn out as completely as his legions: the rest of the people in the cities under his rule were so horribly wasted by famine and pestilence that a single measure of wheat fetched 2,500 Attic drachmas. Hundreds were dying in the cities, still more in the country villages, so that the rural registers which once contained so many names now suffered almost complete obliteration; for at one stroke food shortage and epidemic disease destroyed nearly all the inhabitants.[26]

Without a concept of private charity, no activity was undertaken by individuals, philanthropic organizations, or temples to ameliorate the condition of the sick, and they and their families were left to fend for themselves, often with wholly inadequate resources. "Simply put," writes Rodney Stark, "pagan cults were not able to get people to *do* much of anything. . . . And at the bottom of this weakness is the inability of nonexclusive faiths to generate *belonging*."[27] It was the Christian belief in personal and corporate philanthropy as an outworking of Christian concepts of *agape* and the inherent worth of individuals who bore God's image that introduced into the classical world the concept of social responsibility in treating epidemic disease.[28] The only care of the sick and dying during the epidemic of 312–13 was provided by Christian churches, who even hired grave diggers to bury the dead that lay in the streets. One finds a similar situation more than half a century later in Edessa, in 373, when, during a famine, Ephraem of Syria (c. A.D. 306–73), a deacon, took his own initiative in setting up some three hundred beds in public porticoes for the treatment of the ill.[29] Some beds were for those awaiting burial, and others were for the poor and for strangers. Ephraem's reputation stood so high that he was the only person in the city to whom the rich would entrust their gifts to meet the emergency. He died a month later from ministering to the victims. Again, in about 500, when Edessa was suffering from famine and plague, Christians created temporary shelter in stoas, baths, and other public places.[30]

Earlier, during the plague of Cyprian, Christian churches, even though they were undergoing their first large-scale persecution, devised in several cities a program for the systematic care of the sick. In the autumn of 249 the emperor Decius had ordered senior members of the clergy arrested, and a few months later he required everyone in the empire to offer sacrifice to the gods on pain of death if they refused. In spite of the constraints of persecution, the bishops provided energetic leadership in organizing the clergy to direct relief efforts for those suffering from the plague. In Alexandria, where the plague in a decade reduced the population by more than half,[31] Dionysius, bishop of the city from A.D. 247 to 264, writes that presbyters, deacons, and laymen took charge of the treatment of the sick, ignoring the danger to their own lives.[32] As a result, he writes, "the best of our brothers" succumbed to the disease. Their activity contrasted with that of the pagans, who deserted the sick or threw the bodies of the dead out into the streets.

Further evidence for the Christian response comes from Carthage, where Cyprian was bishop. The plague beset the city in 252, where it caused much havoc. The streets were filled with corpses of the dead, which people were afraid to touch. The pagans deserted their dead and dying, while the unscrupulous took advantage of the situation to rob the sick.[33] The Christians were blamed by the pagans for the calamity.[34] This was a common pagan response, as Tertullian's well-known remark illustrates: "If the Tiber floods the town or the Nile fails to flood the fields, if the sky stands still or the earth moves, if famine, if plague, the first reaction is 'Christians to the lion!' "[35] Cyprian responded to the crisis in an address to the Christian community in which he called on Christians to aid their persecutors and to undertake the systematic care of the sick throughout the whole city. He appealed to rich and poor alike for help. The rich gave of their substance, while the poor were called upon for service. He urged that no distinction be made in ministering to both Christians and pagans. His activity in organizing the care of victims of the plague lasted until his exile five years later.[36]

Although our sources emphasize the voluntary work of the clergy and laity, it is likely that the ferocity of the plague and the high mortality that it induced forced some churches, perhaps for the first time, to employ burial (and perhaps medical) attendants to assist the presbyters and deacons.[37] Gregory of Nyssa, describing the same plague in Pontus, says that "more died than survived, and not enough people were left to bury the dead."[38] We can infer, from the figures given by Dionysius for Alexandria, that the situation there was not much different.[39] According to Dionysius,[40] the Christians undertook the burial of the dead, a task that the pagans refused for fear of contagion. The church had always provided its own members with burial, initially as a work of mercy undertaken by fellow members of the church.[41] The burial of victims of the plague may have seemed to Christian leaders a logical extension of the church's duty to the Christian dead. Christians performed a similar service in a plague that ravaged the Eastern Empire in 312. "All day long," writes Eusebius of that plague, "some continued without rest to tend the dying and bury them—the number was immense, and there was no one to see to them."[42] It is unlikely that so enormous a task in either plague was performed by voluntary labor. Perhaps the staffs of hired grave diggers or sextons (*copiatae, fossores, decani*) that came to be employed by many churches in the fourth century originated during the midcentury plague. By 302 grave diggers had come to be organized as a minor ecclesiastical order in North Africa, where the church at Cirta employed six.[43] The patriarch of Antioch maintained a similar group known as the *lecticarii,* who buried the bodies of the poor, while in Rome the bishop employed *fossores* to tunnel out the rock beneath the city to bury the Christian dead in the catacombs.[44] The Christian churches had become so identified with the burial of the dead by the fourth century that Constantine inaugurated free burial services under the direction of the clergy.[45] Julian the Apostate singled out for mention (along with their hospitality and purity of life) the Christians' concern for proper burial of the dead as a factor that had led to the Christianization of the empire.[46]

It is tempting as well to see in the plague of Cyprian, as did Edward Gibbon,[47] the origin of the medical corps known as the *parabalani* (or *parabolanoi*), which gained fame in Alexandria in the fifth century.[48] They are mentioned in the Theodosian Code (16.2.42/416 and 43/418), where they are said to be entrusted with the care of the sick *(qui ad curanda debilium aegra corpora deputantur)*. They were apparently enlisted from the poorer classes of Alexandria, forming a corps of ambulanciers who were engaged in transporting and nursing the sick. Their number was large (some five hundred men), and they were under the jurisdiction of the patriarch of Alexandria. The *parabalani* developed a reputation for promoting organized violence in the religious and political controversies in Alexandria.[49] In the quarrel between Cyril, bishop of Alexandria, and the Roman prefect Orestes in 416, they terrorized the city in support of Cyril, and in the course of rioting the distinguished pagan philosopher Hypatia was murdered by a Christian mob.[50] As a result, restrictions were placed on their activities by the emperor Theodosius II (401–50). Some of these restrictions were later removed, but the order's propensity for violence continued, and the *parabalani* appeared again at the Latrocinium, or "Robbers' Council," at Ephesus in 449, where they coerced their opponents.

The etymological derivation of their name has been much debated. A widely accepted view connects the term with *paraballesthai* (sc. *tēn zōēn, psuchēn*). They were the "reckless ones," so called from the courage with which they risked their lives in aiding the sick. The establishment of the *parabalani* has generally been attributed to the period after the legalization of Christianity,[51] but if the suggested etymology is correct, it furnishes an argument in favor of their having originated in a time of plague, during which they risked extraordinary exposure to contagion. Eusebius mentions Christians who in many cities in the East, during the plague in 312, performed tasks similar to those performed at a later date by the *parabalani*. Besides those who buried the dead, he writes, "others rounded up the huge number who had been reduced to scarecrows all over the city and distributed loaves to them all."[52] Whether he is describing a voluntary effort or one that employed hired medical attendants like the *parabalani*, we cannot say. Eusebius places emphasis on the large number of both the sick and dying in the plague of 312, and that fact lends credence to the view that a corps of men was hired to transport and care for them. The *parabalani* are known only from Alexandria, and the most likely historical context for the origin of the order is the plague of the mid-third century. Perhaps Dionysius, to provide simple care for the extraordinarily large numbers of the sick, used unemployed men to carry out benevolent work that had outgrown the voluntary resources of the church. The creation of a corps of ambulance personnel would have been intended, of course, as a temporary expedient. However, even if there was no direct continuity, what had been intended as an improvisation may have been a precursor to later organizations of medical attendants like the *parabalani* in Alexandria.

Although the work of the large urban churches during the plague was done on an ad hoc basis, it probably would not have been so effective had not a system of parochial care of the sick already existed. Indeed, the importance attached to voluntary benevolence by the early church obscures the high degree of organization developed in the pre-Constantinian period. And the genius of the church in adapting itself to the increasing demands for its charitable activities is nowhere more evident than in its concern for the poor and the sick. Even in its earliest stages the church's success in caring for the sick depended as much on the carefully defined duties of its leaders as on lay involvement. Christian medical philanthropy furnished palliative care, which lay within the ability of those without medical training to provide. But one should not draw the lines too distinctly, since therapy must have been administered when necessary or available. Rodney Stark argues that, because the pagan worldview had no concept of social service and community solidarity, "when disasters struck, the Christians were better able to cope, and this resulted in *substantially higher rates of survival*."[53] The palliative care that they offered the sick, even the simple provision of food and water, without skilled medical attention, would have reduced mortality considerably. "Modern medical experts," he writes, "believe that conscientious nursing *without any medications* could cut the mortality rate by two-thirds or even more."[54] No charitable care of any kind, public or private, existed apart from Christian diaconal care because there was no religious, philosophical, or social basis for it. Not only did substantial numbers of Christians survive, but since nursing care was given to pagans as well, gratitude likely had a powerful effect on public attitudes to Christianity. The number of Christians increased during the plague as a result of the decline of traditional social bonds and the creation of new bonds between surviving pagans and Christians, resulting in large numbers of conversions.[55]

The diaconal model of philanthropy was well suited to the first three centuries of Christianity, when the urban congregation was the focal point of the movement. I suggest that it was the great plague of the mid-third century that provided the church with its greatest opportunity for the broad extension of medical charity. Its ministry to the sick had hitherto been inwardly directed, largely to its own adherents. Increasingly Christian medical care became outwardly focused, now enlarged to include many who were victims of the plague. The administrative structure was already in place. Deacons, aided by men in minor clerical orders, routinely cared for the sick on a large scale, while presbyters and bishops were experienced in the administration of sizable funds from the collection and distribution of alms. Whether or not they made use of a corps of hired medical and burial attendants,[56] the energetic response of the bishops to the plague marked, I believe, a significant advance in the church's concept of medical charity. The evidence suggests that for the first time the church conceived of its ministry to the sick as one that included both pagans and Christians without distinction.

As late as the mid-fourth century the concept of being a "lover of the poor" (*philoptôchos*) was a novel one in the Graeco-Roman world, with no antecedents in classical ideas of philanthropy.[57] Organized care of the poor ran contrary to patterns of civic beneficence, in which aid was distributed by public benefactors (*euergetai*) to all citizens alike without regard to wealth or status. Within the traditional classical pattern of euergetism, the rich showed their civic patriotism to the city by sharing their wealth, not with the poor but with their fellow citizens. When the sense of community within the city-states was weakened in late antiquity, the old ideological basis for euergetism was replaced by a new ideology of private charity in which one group within society was elevated above the rest as recipients of philanthropy. The introduction of new ideas of almsgiving, which had their origin in ancient Near Eastern (Jewish and Christian) rather than Graeco-Roman values, led to a redefinition of the poor in Christian terms.[58] A specific group defined as "the poor" had not previously existed in the public eye as long as the community was viewed as a collective whole, one in which all citizens of the city (but not outsiders) shared in public benefactions.[59] Evelyne Patlagean argues that this abandonment of the civic model of beneficence for a more narrowly defined one was accompanied by the vast growth of homeless poor that began to crowd into the cities from the countryside of the Eastern provinces, producing a crisis after 450.[60] Peter Brown believes that there was no demographic crisis; the poor were not beggars but *déclassés*, who had always existed in classical society but were invisible to the upper classes.[61] Many of the new poor were probably homeless immigrants, though some were merely people who found themselves in straitened circumstances. Their number included distinguished refugees, but if they were not citizens, they had no status within the city.[62] Wealthy pagans naturally continued to espouse the traditional classical view that the poor were passive recipients of fate, and they looked down on them as base and ignoble in character. Christians, influenced by many biblical texts that spoke of the care of the poor as a duty, saw them instead as especially blessed by God, endued with special grace, and still bearing the *imago Dei.* They regarded giving to them as giving to Christ, and philanthropy to the poor as demonstrating their love for their Savior. Both donor and recipient came to regard themselves as fellow servants, a theme that one finds repeatedly in contemporary sermons.[63] Hence distinctive Christian ideas of charity, which had not enjoyed public recognition until the mid-fourth century, for the first time in classical society both identified and elevated the previously invisible poor as a specific group.

For the Roman government the Christian emphasis on caring for the poor became a defining factor in assigning a recognizable role in Roman society to the church.[64] The state had no interest in caring for the poor and after A.D. 313 was glad to assign it to the church in return for favors granted to it.[65] Richard Finn has demonstrated that the central place of almsgiving in Christian charity and the rapid growth of Christian churches after legalization, particularly in metropolitan areas, led to the increasing power of the bishops. Constantine and subsequent emperors provided large gifts of grain to bishops to distribute and occasionally granted direct subventions.[66] As a result churches began to take over some of the distribution of funds and aspects of the grain dole (*annona*) that had traditionally been a role of the Roman imperial government.[67] We have only a few hints regarding the number they cared for. John Chrysostom estimated in a sermon that one-tenth of the population of Antioch could be considered as belonging to the poor, a statistic that Peter Brown regards as convincing.[68] The administration of episcopal charity was a key factor in the prestige and power that the bishops came to enjoy in their communities. Because Constantine provided food and clothing to the bishops for distribution among the poor in their churches,[69] they were increasingly looked to as the source of charity in their dioceses and, within the larger community, as brokers in gaining protection and influence with imperial authorities. The mingling of the divergent interests of church and state led to "the Christianization of euergetism," in which the bishops assumed a major role as public spokesmen for the poor, a role that, together with their status as major distributors of charitable funds, increased their prestige and status within the city and even with the imperial government.[70] They preached frequent sermons on the necessity of giving alms[71] and used funds from collections, contributions, and legacies of the rich, together with public monies, to maintain an extensive ministry to the poor in the distribution of gifts and the building of permanent charitable institutions like hospitals.[72] What had once been an important role of the state—the distribution of philanthropic funds—was increasingly in the fourth century taken over by the church in exchange for exemption from taxes. The years from 320 to 420 marked the growth and development of this new role, which reached a fully developed state between 451 and 565.[73] "Love of the poor" became a public virtue that was expected even of emperors (but not of their officials, who had no such expectation imposed on them), while the chief duty of the bishop in the centuries following was thought to be to care for "the poor," a term that came to denote the weak.[74] The lower classes of the city, given a specific identity and defined for the first time as collectively deserving the assistance that had previously belonged to all citizens, the *dēmos,* came over time to replace the *dēmos.*[75] This little-noticed movement marks one of the truly revolutionary changes in human sentiment in Western history and constitutes a significant feature of the transition from a classical to a Christian society.[76]

The Origin of the Hospital

The concept of the church's care of "the poor" was basic to the founding of the earliest hospitals. The hospital was, in origin and conception, a distinctively Christian institution, rooted in Christian concepts of charity and philanthropy.[77] There were no pre-Christian institutions in the ancient world that served the purpose that Christian hospitals were created to serve, that is, offering charitable aid, particularly health care, to those in need.[78] None of the provisions for health care in classical times that have been suggested as early exemplars—military and slave infirmaries (*valetudinaria*), temples of Asclepius (*asclepieia*), physicians' clinics (*iatreia*), or public physicians (*archiatri*)—resembled hospitals as they developed in the late fourth century. Roman infirmaries, called *valetudinaria*, which were maintained by legions and large slaveholders, have most often been adduced as parallels or precursors. But they offered medical aid to a restricted population (soldiers and slaves) and were never available to the public. Moreover, they were created for economic or military reasons, not as charitable foundations.[79]

Cenobitic or community monasticism had from its beginning placed a premium on practical charity of all kinds, particularly medical charity, and the rise of charitable foundations occurred in tandem with the growth of the monastic movement. Thus the poorhouse (*ptōchotropheion, ptocheion*), which appeared in the early 340s in Constantinople and elsewhere, accepted the sick as well as the poor. In the early 380s (but perhaps as early as the 330s[80]) hostels (*xenones*) that cared for the sick were attached to churches in the capital city.[81] Separate institutions were established for orphans (*orphanotropheia*), foundlings (*brephotropheia*), the aged (*gerontokomeia*), lepers (*keluphokomeia*), and poor travelers (*xenodocheia*). It was not till the late fourth century that Christian hospitals began to be organized. They were known by a variety of names (*nosokomeia, xenones*) that came to distinguish them as hospitals.[82]

"To attend the birth of the hospital," writes Vivian Nutton, "depends ultimately on a question of verbal definition."[83] Andrew Crislip identifies three necessary characteristics: inpatient facilities, professional medical care for patients, and charitable care.[84] All three were found in the best-known, and probably the earliest, hospital, the Basileias, begun about 369 and completed by about 372 by Basil the Great, who was to become bishop of Caesarea (modern Kayseri), in Cappadocia (modern Turkey).[85] Basil's idea of creating a hospital (or, as Basil himself termed it, a poorhouse [*ptōchotropheion*]) grew initially out of a famine in 368/69 (or perhaps in 369/70), during which he had organized a distribution of food.[86] His hospital, which he established outside Caesarea, employed a regular live-in medical staff who provided not only Christian aid to the sick but also medical care in the tradition of secular Graeco-Roman medicine.[87] It included a separate section for each of six groups: the poor, the homeless and strangers, orphans and foundlings, lepers, the aged and infirm, and the sick.[88] The *keluphokomeia* housed lepers, who were gathered together from the countryside around Caesarea into one place where they could be cared for. Gregory of Nazianzus has left us a contemporary, if somewhat idealized, description of the Basileias, in which he contrasts the treatment received by the sick (particularly lepers, the "*ptōchoi* par excellence"[89]) with their previous condition. Gregory describes it as "a new city [kainēpolis], the treasure-house of godliness . . . in which disease is investigated and sympathy proved. . . . We have no longer to look on the fearful and pitiable sight of men like corpses before death, with the greater part of their limbs dead, driven from cities, from dwellings, from public places, from water courses. Basil it was more than anyone who persuaded those who are men not to scorn men, nor to dishonor Christ the head of all by their inhumanity towards human beings."[90]

Given the fact that no parallels exist in the classical world, the extent to which the hospital is related to its precursors has been much debated. Andrew Crislip argues that the monastic infirmary provided a template for the earliest hospitals that arose in the 370s.[91] What he describes as "an innovative type of health care system" grew up within early monasteries, which incorporated medical treatment, professional attendants, and an infirmary, the last forming a "protohospital" that existed as early as 324, when Pachomius created the first monastic infirmary. The infirmary pioneered all the services that were later included in the Basileias, particularly medical care without charge and inpatient treatment, and it intentionally destigmatized illness. "The similarities between the monastic health care system and the late antique hospital," writes Crislip, "are too great to ignore, and without a doubt the historical origin of the hospital lies precisely here."[92]

The suggestion that the monastic infirmary served as the model of Basil's hospital does not solve the problem, however, but merely takes the question of origins back one step. Crislip writes: "We search monastic literature in vain for an answer to the question of developmental origins. Monastic sources do not describe any development of their health care system. They identify no key event, no watershed moment, no prime mover behind its foundation, nor any step-by-step process in which its structure was elaborated as an entity in itself. Rather, monastic sources take their health care system as a given, as an integral part of monastic life from its inception."[93] The answer lies in part, as Crislip points out, in the fact that monks, in renouncing the world, abandoned ties with family and property and hence with outside support and social aid.[94] It became necessary to create within the walls of the monastery both a surrogate family and an alternative social system for monastics who had left their own families behind. But Crislip underestimates (though he acknowledges) another important factor, namely, that the experience gained by the congregation-centered care of the sick over several centuries gave early Christians the ability to create rapidly in the late fourth century a network of efficiently functioning institutions that offered charitable medical care, first in monastic infirmaries and later in the hospital.[95] So too does Peter Brown, who argues that it was the conversion of Constantine that "dramatically altered the scale of Christian charity, the nature of its institutions, and the meaning that such charity took on for a still partially christianized world. It was no longer a fiercely inward-looking matter, directed to the needs of the poor."[96] But the church's medical charity had ceased to be inward-looking a half century before Constantine. When the concept of a hospital began to emerge in the mid-fourth century, it owed much to the church's long experience in caring for the ill and to its careful attention to the organization of charity within a congregation-centered pattern. Both were legacies of the first three centuries of Christianity, and without them the immediate success of the hospital, I believe, would have been impossible.

As suggestive as Crislip's theory is, it contains an element of Whiggism. Just as the Pantocrator hosptial, founded in 1136, represented the culmination of the development of the Byzantine hospital and in some reconstructions of hospital history casts its retrospective shadow over the early history of the institution,[97] so the Basileias in Caesarea plays a similar role as the greatest of the early Christian hospitals. Of course, Basil's hospital was not the culmination of a long period of development, since it appears to have been the first hospital founded. But Crislip sees the end from the beginning: He views the history of the hospital as moving progressively from the infirmary in Pachomius's monastery to the Basileias, an institution that defines for the historian what constituted the early Christian hospital. I suggest, however, that the institutional situation is both more complex and more multiform. Beginning in the 320s there arose a number of discrete institutions that were devoted to a single purpose: hostels for the convenience of travelers, orphanages for the care of foundlings, homes for the elderly, and almshouses for the poor. In many of them some form of physical care was given to those in need, and it would be surprising if nonprofessional medical care was not made available when circumstances necessitated it. The "development" of the hospital did not cease with the Basileias, however. Many of the more specialized institutions continued their existence long after Basil's foundation. While serving as a model for later hospitals, it did not immediately change or broaden other charitable institutions. Peregrine Horden's words, though describing the Pantocrator, are pertinent here: "If we want to identify the golden age, we must turn away from the small number of grandiose edifices Byzantium produced. We must, in effect, lower our sights—to the humbler, but more numerous hospitals, hostels, almshouses and orphanages which perhaps made up in numbers what they lacked individually in resources and personnel. Effective philanthropy need not always depend on a complex administration, massive funding and a detailed instruction manual."[98] If we stress the gap that separated the protohospitals that preceded the Basileias from the first "fully developed hospital," we court the danger of imposing essentialist definitions on the development of the institution.

In a programmatic essay Horden suggests several ways in which we can progress beyond what he terms "an elementary positivism" in dealing with Byzantine welfare, particularly in the realm of medical philanthropy. The first is that we "rid our typologies of anachronistic notions of the difference between 'caring' and 'curing'—as if only a well-defined medical profession (such as did not exist in Byzantium) were capable of the latter."[99] The practice of ancient medicine cannot be easily reconciled with modern ideas of medical professionalism, a factor that pertains to the distinction that is often made between caring (delivering palliative care) and curing (providing medical therapies). In a society that lacked both medical licensure, with its restrictions on who could practice medicine, and any defined nursing profession, the boundary was fluid. While physicians applied therapies that were informed by Galenic medicine, their therapies may not have differed appreciably from those of monks who were well informed about medical theory and skilled in offering care. Hence one might be hard pressed in certain instances to distinguish the treatment provided by an experienced caregiver from that prescribed by a physician.

In assessing the place of the Basileias in the history of hospitals, one needs to ask what was novel in Basil's creation. The charitable aspect was not, since other institutions of a more limited nature had already been established; it has been suggested that Basil's mentor, Eustathius of Pontus, may have influenced his ideas of poor relief.[100] None of the areas into which the Basileias was divided were new. The sick, lepers, the poor, travelers, orphans, and the elderly were already being cared for in more specialized institutions. The rescue and care of orphans and foundlings was regarded by early Christians as a particularly Christian duty, since it involved in many cases saving the lives of children who had been exposed by their parents.[101] Because the exposure of newborn infants was a widespread feature of pagan society,[102] the number of foundlings was large, and Christians began in the fourth century to develop orphanages (*orphanotropheia*) for their care. What *are* novel are the nonprofessional staff of doctors and medical attendants, the offering of inpatient care, and the comprehensive nature of the institution. But while its size and extent made it unique for its time, it was not an exclusively medical institution. There is no question that it marks a major advance in medical care, and perhaps—if Crislip's thesis is correct—the Basileias may not have come into existence without the model of the monastic infirmary. The "fully developed hospital" did not, however, immediately replace Christian charitable foundations of a more limited kind. There continued to exist a wide variety of such institutions. One of the reasons, surely, is that in their creation much depended on the initiative of a bishop or an abbot. Local circumstances, the extent of the founder's vision, and the availability of funds and land were determining (and limiting) factors. Moreover, as Peter Brown reminds us, the founding of a hospital fit nicely into the traditional role of the *euergetes.* "It was both a work of public, civic munificence and an act of public charity."[103] It brought with it further tax exemptions and magnified the role of the bishop who founded it, particularly if it included a notable building.

Horden suggests that we need to consider the question of patients' demand for hospitals in "an analysis which acknowledges the vitality of the 'make-shift' economy of the poor, the smallness of most centers of settlement, and the possible undesirability of entering a hospital." The present is liable to deceive us, living as we do in a society in which hospitals are readily available. They were not so in the ancient world, where vacant beds might not be available. Counting hospitals and even beds is not enough, warns Horden, especially in late antiquity, when the pressure on hospitals must have been great, perhaps greater than in the later centuries of Byzantium.[104] Like Evelyne Patlagean, Andrew Crislip believes that in the fourth century a large immigration of rural poor took place into the cities of the eastern Mediterranean and that an urgent demand arose for the treatment of the sick. Hence the Christian charitable institutions met an immediate social and medical need.[105] Vivian Nutton observes how small some of the institutions were for rather large cities and how small a proportion of the sick they could house.[106] Finally, Horden reminds us that the medical treatment offered in hospitals was only a part of the larger medical world, which consisted of a variety of practitioners, healers, and holy men. We cannot assume that the chronically or seriously ill would have considered the hospital the healer of first resort. In fact, just the reverse is likely to have been the case: most patients continued to be treated in their homes by physicians long after the genesis of the hospital.[107]

Historians of Christian medical philanthropy have stressed the importance to health care, both short term and long term, of the creation of the hospital. Timothy Miller has underscored the rapid spread of the institution,[108] and Vivian Nutton has remarked on the change of attitude that, "within a century, set the hospital in the front line of defense against illness."[109] One's evaluation is a matter of perspective, of course, but by focusing on the hospital created by Basil and the rapid growth of charitable institutions of care and healing, modern historians have not escaped Whiggish approaches. The point is illustrated by the spectrum of definitions given to the word *hospital,* which continued to be fluid in late antiquity. Thus Nutton speaks of the erection of "hospitals" for the victims of famine and plague in Edessa in 500/501,[110] when, in fact, they were temporary shelters in stoas, baths, and other public places.[111] These emergency shelters will hardly fit Crislip's definition of hospitals, hastily constructed as they were in an ad hoc manner; yet they were founded well over a century *after* the permanent institution created by Basil, to which they owe little or nothing. They were, however, intended to meet an immediate need. They tell us something of late antique medical philanthropy that the Basileias does not, and they are as firmly rooted as it is in Christian concepts of charity.

Basil played a pioneering role in bringing monasteries within the administrative structure of the church. He brought lavra and cenobitic monasticism into a close relationship, created a new role for them in emphasizing practical service, and undertook an ambitious new program of institutional charity within the monasteries. He secured funds for his initiatives, and wealthy individuals came to play an increasing role in establishing hospitals.[112] Municipal bishops long exercised a general supervision of charitable institutions, many of which they founded, and supported them with ecclesiastical funds.[113] Later they came to enjoy the largess of emperors.[114] Hospitals quickly expanded throughout the Eastern Empire in the late fourth and fifth centuries, with bishops taking the initiative in founding them.[115] They spread to the Western Empire a generation after they were founded in the East, but owing to economic difficulties, their growth in the West was much slower. The earliest Western hospital was established in Rome around 390 by Fabiola, a remarkable and independent-minded noblewoman who was a friend of Jerome's. Jerome writes, doubtless with some exaggeration, that the hospital in Rome enjoyed such success that within a year after its founding it was known from Parthia to Britain. Fabiola built the hospital with her own funds and worked in it herself, gathering the poor sick from public squares and personally nursing many of them. Her own participation (like that of Basil) was a factor that distinguished Christian charity from euergetistic philanthropy.[116] Hospitals and other charitable institutions were recognized as peculiarly Christian institutions, and the emperor Julian (A.D. 360–63) complained that "the impious Galilaeans support not only their own poor but ours as well; everyone can see that our people lack aid from us." In 362 he urged, in a letter to Arsacius, the chief priest of Galatia, that pagan charitable foundations be established in every city for those in need, both for their own people and for foreigners.[117] The tone of the letter makes it clear that the request is intended to recapture the initiative from the Christians rather than inspired by personal philanthropic motives. Although his comments are based on the institutions that predated the Basileias, his intention was to emulate the Christians.

Given the wide range of specialized early Christian charitable institutions that went by the name of *xenodocheia,* not all cared for the sick, and only a minority of even them had the resources to employ physicians. Horden estimates that in the pre-1204 period some twenty-three to twenty-five Byzantine hospitals had physicians.[118] More commonly employed were *hypourgoi,* assistants who had no particular medical training.[119] It would be an anachronism to speak of a professional hospital "staff." It is for the most part in Byzantium that one finds physicians at all; in western Europe, except in Italy, there were few physicians until the end of the Middle Ages.[120] The medicine administered was at a low level, again given the limited facilities available.[121] But hospitals cared for the soul as well as the body. The attention given to the healing of the soul in later, Western medieval hospitals, based on an understanding of the healthy soul's contribution to the health of the body, has been described as psychosomatic medicine. In the tradition of Christus medicus, administering spiritual medicine was the first duty of medieval hospitals. Caregivers were aware of the importance of rest, diet, and nursing care, but they recognized that the "passions of the soul" were important in healing and especially encouraged cheerfulness.[122] Basil considered psalmody important in soothing the soul.[123]

Earlier hospitals—those of late antiquity—grew out of the monastic movement, and the widespread existence of monastic orders provided much of the personnel to staff medical institutions. In many cases the model of earlier, palliative care of the sick remained the only care available. Over time some hospitals (always a minority) came to employ physicians. The entry of numbers of Christians into medicine in the late fourth century may have been motivated in part by the desire to serve the ill in hospitals.[124] The first hospitals were founded to provide care for the poor. The pattern persisted, and hospitals remained for centuries what they had been intended to be from the beginning, institutions for the indigent (although they provided other medical assistance), while those who could afford a physician's care received it in their homes.[125]

Spoudaioi and *Philoponoi*

Not as well known as the *parabalani,* but more widespread, were the lay orders of *spoudaioi* and *philoponoi,* which were to be found in the cities of the Eastern Roman Empire in late antiquity and the Byzantine period. The *spoudaioi* were "the zealous ones"; in Egypt they were known as *philoponoi,* "lovers of labor." They formed groups that were attached to large churches in the great cities of the East: Alexandria, Antioch, Constantinople, Beirut, and Jerusalem, most prominently, although they are attested for smaller cities as well.[126] The *spoudaioi* comprised both lay men and women who adopted ascetic practices that included chastity (or continence for those who were married) and fasting. Their functions varied somewhat according to local traditions, but in most cities their ministry was twofold: to care for the sick and to perform liturgical functions, such as reading scripture, chanting, praying, and participating in funerals, vigils, and processions.[127] While they represented a branch of the ascetic movement, they never constituted a monastic order and so stood apart from both the anchoritic and cenobitic traditions.[128] Although laypeople, they came to be recognized as an intermediate order between the clergy and the laity, and they are so described by several sources.[129] Our earliest reference to the *spoudaioi* is found in a letter written in the year 312 by the patriarch of Alexandria, who mentions them in the context of the years 303–5. We find frequent mention of them from the fourth through the seventh centuries and a few scattered references thereafter.[130]

Our sources indicate that a chief function of the *spoudaioi* was to provide assistance to the indigent sick of the urban areas in which they lived. Already in the Hippocratic Corpus we find reference to the homeless sick who populated the streets of Greek cities in the fifth century B.C. While widely removed in time from late antiquity, they provide a vivid picture of social conditions that did not change in essentials over many centuries.[131] Several cases are recorded in the *Epidemics* (which was probably written about 400 B.C.): a girl in Abdera, who lay sick by the Sacred Way for twenty-seven days; Anaxion of Abdera, who lay sick of acute fever by the Thracian Gate for thirty-four days; the wife of Delearces in Thasos, who lay sick on the plain with acute fever and mental disturbance.[132] In many instances, as one often finds in the *Epidemics*, those whose conditions are described died after a specified number of days. The fact that they remained uncared for in public places suggests that they were without resources and were either set out to die or had no family or friends to care for them. A well-known biblical example is the narrative in the Fourth Gospel (Jn. 5:3–4) of Jesus's healing of a man who had been paralyzed for thirty-eight years. He lay at the pool of Bethesda, situated near the Sheep Gate in Jerusalem, which had five porches around it. Those who were blind, lame, or paralyzed gathered there, waiting for the water to be stirred up, since a local tradition attributed healing properties to the movement of the water. The man must have been without family or close friends because he is reported to have said that he had no one to bring him to the pool when it was stirred up. The natural inference is that he was homeless. One finds similar pictures of the poor and disabled who congregated in public places in late antiquity as well: a woman lying in labor in a church portico at midnight; the poor seeking warmth in the public baths on winter nights.[133] The picture was a familiar one reflecting, argues Peter Brown, not time-specific events that reflected a declining Roman Empire but the kind of poverty that had always existed in the Mediterranean world, depicted in sermonic literature in conventional tones to elicit sympathy. Brown describes in some detail how broad the Christian definition of poverty was in practice.[134] It involved the care of orphans and widows, not always impoverished but in danger of becoming so ("Distressed Gentlefolk"), as well as the destitute poor. What was new was that it was noticed for the first time by Christians, who saw the poor as a discrete group who needed assistance.[135]

In classical antiquity the household or family[136] (Gk., *oikos*; L., *familia, domus*) provided the chief locus of health care.[137] The family has been described as "the only ancient safety net of real importance."[138] The two chief alternative sources to family care—namely, the patronage system, such as the Roman patron-client relationship, and public philanthropy—made no provision for the health care of the destitute.[139] It was not uncommon for the chronically ill to be shunned, either because they posed too great an economic burden on a family whose very survival was threatened or because of the risk of contagion. Slaves were often abandoned on the Tiber Island at Rome, where they could seek healing when they were too old or too sick to be profitable, and the elderly (people over 60) were sometimes left to fend for themselves, especially if they were unfortunate enough to outlive other, younger family members, a not uncommon occurrence.[140] Individuals who were without family to provide care were in a precarious position when it came to finding food, clothing, shelter, and health care. There was no provision in Graeco-Roman society for public or private shelter or care of any kind for those who were destitute. Hence they were often forced to live on the streets, or in porches, tombs, or makeshift dwellings. Public baths provided fresh water that was essential for hygiene (physicians prescribed hydrotherapy for many specific diseases[141]) and furnished some warmth in cold winters. Some of the poor sought the assistance of Asclepius in Asclepiea. An occasional sick person mentioned in the *Epidemics* is said to have received help from a physician. How commonly it occurred we cannot say, but many in the ancient world suffered from chronic or disabling conditions for which there was no cure. Those afflicted with mental disorders or loathsome diseases were often driven away, as we see recorded in several instances in the Gospels.[142] Even in time of plague no public services were maintained by municipalities to bury the dead, who were thrown out onto the streets.

In a poem about the hardships of a beggar's life, Martial (10.5.11 ff.) depicts a derelict man in his dying moments listening to dogs howling in anticipation "of eating his corpse; at the same time he tries to keep birds of prey (*noxias aves*) at a distance by flapping his rags at them. A gruesome but probably commonplace event in the capital. The poor and destitute, lacking concerned relatives, would be left to rot in the streets, though if Martial's picture is accurate, dogs and vultures would set to work before a cadaver had time to putrify."[143] Here as elsewhere in the classical world, self-help was taken for granted.

It was to these urban poor, sick or dying on the streets, that the *spoudaioi* devoted their service. John Chrysostom described the poor of Antioch, who "wandered around like dogs in the alleys and haunt the corners of the streets . . . they cry from their cellars, calling for charity."[144] The *spoudaioi* would frequently search the streets and alleys by night for those who were ill, distribute money to them, and take them to the baths. Their number must have been large in the major cities of the late Roman Empire, where poverty was ubiquitous.[145] There is no evidence that the *spoudaioi* were skilled professionals or that they formed a separate nursing order. Although they were medical attendants, they had no medical training.[146] The functions that they performed, moreover, were those that would ordinarily be performed only by members of the lowest classes. Timothy Miller argues that *parabalani* and *spoudaioi* brought the disabled poor to hospitals before the sixth century, when, he believes, they were replaced by professional medical officials.[147] There is, in fact, no evidence to associate either order with hospitals. Hospitals arose quite independently of the *spoudaioi* and were based on different principles. From the beginning they offered at least some medical care. Basil's hospital at Caesarea employed medical attendants, and John Chrysostom hired physicians for the hospital for lepers that he founded in Constantinople.[148] In contrast the *spoudaioi* had no medical training, and they were not associated with professional healers.[149] With a few exceptions we do not hear of their administering drugs.[150] In Alexandria the *philoponoi* carried the sick to the church of SS. John and Cyrus, not to a hospital. They presumably employed ad hoc arrangements, approximating those used in time of plague to house the sick.

Hospitals differed from the work of the *spoudaioi* as well in their close relationship with monasticism. The philanthropic impulse of cenobitic monasticism found a natural outlet in the hospital. Beginning with the Basileias, monks came to be involved during the first century of the hospital's development with nearly every hospital in the Eastern Roman Empire.[151] In contrast, the functions of the *spoudaioi* were centered in the great churches of urban areas and were viewed as an extension of those churches' ministry. The *spoudaioi* were the heirs of a long tradition of medical care within a parochial pattern that had its origin in the diaconal care of the sick. That care had grown enormously, but in many respects it had not departed from the original pattern. It was centered in the church, it emphasized lay (as opposed to trained medical) care, and it was under the direct supervision of the bishop. The lay orders of medical attendants continued the much earlier practice, begun in times of plague, of assisting the indigent sick and dying on the streets, a tradition that was only occasionally incorporated into the treatment of hospitals. Fabiola, who founded the first *nosokomeion* in Rome, personally gathered the sick from the streets of the city.[152] In most cases, however, the tradition of transporting the indigent sick to hospitals was carried on by private initiative and was never an integral part of the operation of the hospital.[153]

Although there is no evidence that the *spoudaioi* were in any way associated with the hospitals that began to be founded in the late fourth century, they were responsible for the establishment of an analogous institution, the *diakonia.* In the early Byzantine period the *spoudaioi* were found among both the Chalcedonian and Monophysite parties.[154] One of the most prominent Monophysite leaders of the *spoudaioi* was Paul of Antioch, who later became the Monophysite bishop of Antioch. Paul introduced a distinctive dress for the *spoudaioi* who were under his direction, as well as a hood to hide their faces. He established *diakoniai* in several cities, including Constantinople and its suburbs. According to John of Ephesus, the *diakoniai* based their treatment not on the doctrines of Hippocrates and Galen but on God's word. The *diakoniai* were lay organizations maintained by *philoponoi,* who were not medically trained.[155] John's description of the services they offered makes it clear that the *philoponoi* limited their services to providing palliative care, clothing, bathing, and anointing. Hence their group functions were merely a collective extension of their individual ministry. Why were the *diakoniai* founded when there were already hospitals in existence? The answer is surely that they met a need that the hospitals were not meeting, especially in the care of large numbers of indigents who required personal care.

Some studies of the *spoudaioi* and *philoponoi* have attempted to place them within the context of medieval confraternities by pointing to their resemblance to those in western Europe.[156] While they shared similarities with later medieval analogues, such as the associations of "Poor Men" who stressed humility, poverty, and simplicity, or the "Third Orders" of pious lay men and women established by the mendicant orders, equally significant differences existed. Our sources omit any reference to fraternal convivial occasions like common feasts, christenings, and funerals among the *spoudaioi.* They portray relatively unstructured groups with common duties but little organization and not even a homogeneous background. Perhaps their origin is to be found in groups of men and women who devoted themselves to prayer and acts of mercy in the great churches of the eastern Mediterranean. In some churches their functions became regularized in the fourth century, and some were even established as minor ecclesiastical offices. However, they were usually appointed rather than ordained. In several papyri the *spoudaioi* are included in lists of ecclesiastical officials.[157] In one such papyrus,[158] a list of distributions of wine dating from the sixth or seventh century, probably from Oxyrhynchus, *philoponoi* are mentioned, together with grave diggers and *parabalani.* That the *spoudaioi* remained a minor order, even while consisting of laypeople, is indicated by the latest text to mention them, a list of ecclesiastical officials that dates from the end of the tenth century.[159] One or two minor orders were eventually transformed into guilds. At Constantinople the *dekanoi* (*decani*) were formed into a collegium by Constantine and granted certain privileges and immunities. Their number was fixed at first at 1,100 and later at 950 members. The *parabalani,* after creating a disturbance, had their numbers reduced by Theodosius II to 500 in 416 but increased two years later to 600. After 418 they continued to be under the control of the patriarch, but their status resembled that of a guild.[160] Unlike the *decani* and *parabalani,* the *spoudaioi* were never organized into a guild, and in several respects they differed from guilds. The *parabalani* and *decani* seem to have been employed to perform their duties. While they doubtless began as voluntary organizations, each developed into a paid corps under the supervision of the patriarch before being organized into a guild.

In contrast the *spoudaioi* retained their voluntary identity. They apparently were not paid wages but were supported from the funds of the church. Some supported themselves by their own labor. Their ascetic tendencies suggest that they led humble lives. But while the *decani* and *parabalani* were most likely drawn from the lowest classes, the *spoudaioi* sometimes included members of aristocratic families who dedicated themselves to a life of service to God and the relief of suffering. Although wealthy citizens (*curiales*) were forbidden to join their ranks to escape liturgies (financial obligations imposed by the state), we hear of a steady stream of men and women entering the order who likely came from middle and even upper-income strata.[161] Hence the lay medical orders represented a wider variety of social backgrounds than did the poorer *decani* and *parabalani.* Whether their numbers were anything close to these two orders is unknown. But in the great patriarchal churches of the eastern Mediterranean, the number of *spoudaioi* or *philoponoi* is likely to have been considerable.[162]

While we find few references to the *spoudaioi* and *philoponoi* after the early Byzantine period, enough exist to indicate that they continued for several centuries. Their infrequent mention later may simply be a reflection of our sources, which are plentiful for the early period but scanty after the seventh century.[163] Most scholars believe, however, that the lay orders of medical attendants declined or lapsed after the seventh century.[164] At Jerusalem the patriarch Elias gathered the *spoudaioi* of the Church of the Resurrection into a monastery in the late fifth century. We do not know whether this pattern was repeated elsewhere, but it may well have been. Conditions after the Arab invasions may have been inimical to their continued existence as well.[165]

For some three centuries before the genesis of either the monastic infirmary or the hospital, the early church employed another model of medical philanthropy, the parochial care of the sick, which had its origins in the diaconal ministry of concern for meeting the physical and material needs of those in distress. This earlier model was based on the unskilled care of laypeople who attempted to alleviate sickness; professional medical care was for the most part beyond their competence. This pattern, centered in the church, was established at first on a voluntary basis. Later it employed a staff of minor clerics and finally complemented them in meeting the widespread needs of the urban poor of the Eastern Empire by encouraging lay orders like the *spoudaioi* and *philoponoi.* So firmly established was the lay tradition that it continued to function even after the foundation and rapid spread of hospitals. The latter were staffed by monks and—when resources permitted—physicians (often one and the same), while the earlier tradition was maintained by laypeople without medical training. Hospitals and the lay orders coexisted, I suggest, as complementary models of the church's care of the sick. The lay orders provided services beyond those that the hospital could offer: they actively sought out the urban poor who were scattered in public places and who might find access to hospitals difficult.[166] They flourished where there were large churches that could sustain them. They were never paid like the *parabalani* and *decani* or developed a guildlike structure. They generally (though not everywhere) retained their lay character, with mixed philanthropic and liturgical functions, as a "third order" between clergy and laity. There is little evidence that they displayed any hostility to medicine, as has sometimes been suggested.[167] They simply carried on the practice long established in Christian churches of offering the kind of assistance to the sick that laypeople could perform. It was not on the hospitals, as Timothy Miller has suggested, but on the lay orders of medical attendants that the mantle of the early church's diaconal care of the sick fell in the early Byzantine period.

Asclepius and Christ

"The quest for health," writes Shirley Jackson Case, "was one of the most urgent personal demands made upon the deities by people living in the Roman Empire."[168] Most religions of the classical world included an element of religious healing, which was intended to complement rather than to compete with secular medicine. In the Roman Empire those who desired supernatural healing could seek help from a variety of gods, goddesses, demigods, and heroes.[169] Undoubtedly, the most important was Asclepius, whose cult had spread throughout the Greek world before it was brought to Rome in 291 B.C., where the god came to be called Aesculapius (although the variant is much earlier). By the second century he had become the healing god par excellence, who was worshiped either alone or in conjunction with other gods at 732 temples or shrines, 670 of them in the Mediterranean world.[170] These shrines were not merely centers of worship but sites to which pilgrims came for healing, much as they travel today to Lourdes or Fatima. Healing at the temples of Asclepius was sought chiefly by means of incubation, which was the practice of sleeping in the sacred precinct, where, it was believed, the god would appear in a dream to effect a cure or offer a remedy.

As the most common form of divine healing in the classical world, incubation was a feature of a number of cults besides that of Asclepius, including the oriental mystery religions.[171] By the end of the second century many gods, both Greek (such as Hygieia and Pan) and Eastern (such as Isis and Serapis), employed it. One of the greatest attractions of these deities to potential converts, in fact, was their claim to be able to heal. Their temples complemented already-existing shrines of local heroes and sacred springs that had drawn the sick to nearby sites for centuries.[172] Healing cults advertised their cures by public testimonies or aretalogies, such as the *iamata* that were displayed at the shrines of Asclepius.[173] These aretalogies, though often formulaic, were sufficiently convincing to draw those who could not obtain healing through secular medicine. It has been suggested that healing shrines, particularly those of Asclepius, were popular in Roman times, in part because they offered healing without charge to the poor, who could not afford the services of a physician.[174] But they did not serve the poor alone. One of the most interesting ancient accounts of healing by Asclepius comes from the writings of the Greek rhetorician and hypochondriac Aelius Aristides, who spent much of his life in the pursuit of personal health.[175] He became a devoted servant of Asclepius, whose help, he believed, he had often received.

In spite of superficial resemblances, Christianity differed markedly from the pagan cults of classical antiquity in its approach to healing. Temples of the healing gods, like Asclepius or Serapis, served as focal points for those seeking divine healing for their physical afflictions. As the aretalogies indicate, some received healing, perhaps permanently; others, as in many cases the long term would show, were healed only temporarily. But countless numbers of pilgrims undoubtedly went away disappointed by the failure of the god to heal them. Had their faith not been strong enough? Was their physical disability too great even for the god to heal? Was there some impediment, some failure to perform a vow or offer a sacrifice? We hear only of successes. One suspects, however, that failures were more common. The widespread (indeed, nearly universal) belief in the ancient world that the sick person bore responsibility for his or her illness meant that much self-blame and mental anguish are likely to have accompanied the physical pain of those who had failed to be healed. The Christian church in contrast offered a kind of assistance to the physically afflicted that was both less spectacular and more lasting: the care and relief of sickness and suffering experienced by members of its own community. The Edelsteins argued that Asclepius was an especially philanthropic god, who demonstrated a special concern for the poor that his worshipers did not receive from the more distant and remote Olympian deities. They suggested that the hostels attached to the Asclepieia were precursors of the first Christian hospitals, where those who were too poor to afford the attention of physicians were freely cared for.[176] If the Edelsteins had been correct, they would have a valid basis for comparing the philanthropy of Asclepius to that of early Christians. In fact, however, the evidence for their thesis is very meager.[177] What pilgrims sought from Asclepius was healing. Of long-term care and compassion for suffering demonstrated in the Asclepieia we hear nothing. In stressing the importance of Asclepius as a pagan rival of Christ,[178] the Edelsteins misrepresent the early Christians' ministry to the sick before the late fourth century, which did not compete with Asclepius in claiming to offer miraculous healing. Rather, it established a role, previously unknown in the ancient world, of charitable concern for the sick, which ultimately led to the creation of both *diakoniai* and the earliest hospitals.

Conclusion

Christian charity was fostered in the closely knit community of the early church, which demonstrated its corporate concern practically. The cities of the Roman Empire could be lonely for individuals without a support system. "Such loneliness must have been felt by millions—the urbanised tribesman, the peasant come to town in search of work, the demobilised soldier, the rentier ruined by inflation, and the manumitted slave. For people in that situation membership of a Christian community might be the only way of maintaining their self-respect and giving their life some semblance of meaning. Within the community there was human warmth: some one was interested in them, both here and hereafter."[179] Christians created what has been termed "a miniature welfare state in an empire which for the most part lacked social services."[180] Though it was originally directed almost exclusively to the Christian community, the church's program of caring for the sick reached out in times of plague to its pagan neighbors and was highly effective in making converts.[181] It was modified over time. Centered at first in the largely voluntary diaconal ministry of the local congregation, it was gradually extended by the growth of additional clergy, many of them in minor orders. Later it was enlarged by the employment of hired attendants to meet emergency needs in time of plague.

Neither the pagan temple nor the mystery religions created a caring community similar to that found in the Christian *ecclesia* because both lacked an ideological basis for a program of helping the sick. "Love of one's neighbour is not an exclusively Christian virtue," writes E. R. Dodds, "but in our period the Christians appear to have practised it much more effectively than any other group. The church provided the essentials of social security: it cared for widows and orphans, the old, the unemployed, and the disabled; it provided a burial fund for the poor and a nursing service in time of plague. But even more important, I suspect, than these material benefits was the sense of belonging which the Christian community could give."[182] Dodds suggests that it was the Christians' success in creating a community that cared both for its own and for others that was "a major cause, perhaps the strongest single cause, of the spread of Christianity."[183] The philanthropic motive of the church was essential to its early success, and the church never lost sight of its program of caring for the indigent who suffered physical affliction. Indeed, in its development and extension of that role lies Christianity's chief contribution to health care.

Some Concluding Observations

Modern reconstructions of the attitudes of early Christians to disease and healing have been varied. Some scholars maintain that early Christians believed in a demonic etiology of disease. While on first reading this view seems to gain support from the Gospels' accounts of Jesus's healings, a closer examination indicates that there is little in them to suggest that a theory of demonic causation of ordinary disease was held by either Jesus or the first generation of Christians, even if they believed that demons caused disease on occasion. I have sought to demonstrate that early Christians accepted the same naturalistic assumptions regarding disease that were held by most of their contemporaries, whether Jewish or pagan. Early Christian concepts of disease were those of Greek medicine, which had spread throughout the Mediterranean world during the Hellenistic and late Roman Republican eras and had come to be accepted by the majority of people living within the Roman Empire. In the pages of the New Testament one finds little evidence of miraculous healing except in the Gospels and Acts, where it is attributed exclusively to Jesus and the apostolic circle. With the possible exception of James 5:14–15, which furnishes dubious evidence for a prescriptive rite of prayer for the sick, it is not mentioned in the Epistles. Since they were intended to provide normative teaching for the newly founded churches, its absence argues against its being a significant factor in the first-century life of the church. To the contrary, the evidence, scattered and circumstantial as it is, suggests rather that Christians looked to ordinary means of healing—medicine and folk or traditional remedies (Paul's advice to Timothy to take a little wine for his stomach [1 Tim. 5:23] is an example of the latter). In the fifth century widespread belief in demonic activity, the influence of asceticism, and the tendency of those of all classes to trust in miraculous accounts of healing made popular a variety of forms of religious healing. Even then, however, the evidence suggests that physicians and folk-healing practices remained the choice of first resort by Christians.

In the extant writings of the church fathers we find that, far from distrusting the medical art, nearly all of them praised its utility and efficacy. Those like Tertullian, Origen, Tatian, Marcion, and Arnobius, who some scholars have suggested were ideologically opposed to medicine, on close examination appear to have maintained far more nuanced views. More than anyone else, Tatian might be cited as one who warned Christians not to rely on medical healing. But studies indicate that his real concern was not with the medical art per se but with the use of drugs, which he believed allowed demons to gain access to the body. Even movements considered heretical, whose distinctive theology might have had the potential for encouraging religious healing, appear no different in this regard from more orthodox forms of Christianity. Thus Montanists, who maintained the continuing validity of apostolic gifts (prophecy and glossolalia, in particular), apparently did not claim miraculous healing as one of those gifts. Not until the sixth century do we find evidence of Montanists resorting to miraculous healing that was connected with the tombs of the original prophets of the movement.

One finds nuances within this broad framework of a general acceptance of secular medicine in the early church. Some fathers recommended against the use of medicine in those cases of illness that were sent by God for chastisement. For example, Basil of Caesarea, in his *Long Rule* (55), urges monks to avoid the use of medical means when they believe that God has sent illness to discipline or correct them for some sin they have committed. A smaller number of the fathers, of whom Origen is the best known, believed that while the use of medicine was appropriate for the ordinary Christian, those Christians who sought a higher level of spiritual maturity should rely on prayer alone.[1] Norman Baynes thought this distinction between two classes of Christians to be central to Byzantine civilization: "the double ethic which is of primary significance to East Roman life—two standards: one for the ordinary Christian living his life in the work-a-day world, and the other the standard for those who were haunted by the words of Christ, "If thou wouldst be perfect."[2] This attitude was maintained by Christian ascetics in late antiquity, but it remained a distinctly minority position among Christians generally. In both cases the question was not whether it was right to use medicine of any kind but under what conditions one might legitimately *decline* to use it. Nevertheless, while regarding its practice by physicians and its use by the ill as entirely consistent with Christian virtue, the fathers warned that Christians must not place their faith in the means rather than in God to heal disease. The efficacy of drugs comes from God, they argued, who chooses to work through means but can heal equally well without them. A Christian should pray when taking medicine, not merely after the physician's efforts have failed. It is a leitmotif in the writings of the fathers that it is wrong to disparage medicine but equally wrong to trust in physicians rather than in God. Only a minority of Christians would disagree.

Early Christians were the first in the ancient world to endue sickness with positive value. In the classical world good health (Gk., *hygieia*; L., *salus*) was an essential component of a balanced and controlled personality, both a virtue (*arete*) and an indicator of virtue, and hence the sine qua non of the good life, "the first and best possession."[3] Sextus Empiricus (fl. c. A.D. 200), a physician and philosopher of the skeptical school, writes that to ordinary people health is the summum bonum, the highest good.[4] The Greeks believed that without good health nothing else in life could be enjoyed. "When health is absent," writes Herophilus, "wisdom cannot reveal itself, art cannot become manifest, strength cannot fight, wealth becomes useless, and intelligence cannot be made use of."[5] The prominence that the classical world attributed to health reflected the significance placed on the body and on physical culture in Greece and Rome.[6] In the second century of our era a plethora of works appeared, written by both physicians like Galen and Soranus and laypeople, "who saw in their bodies," writes Crislip,

> an important locus for moral reflection and self-examination—how to care for the body, how to provide for its proper functioning, how to control the passions. A preoccupation with the body and its functions was an accepted part of Roman aristocratic life, as seen in the letters and orations of such second-century dignitaries as Aelius Aristides and Fronto, tutor of Marcus Aurelius. While their constant self-examinations and their graphic discussions of every substance entering or exiting the body strike modern readers as almost pathological, these were normal, accepted features of Roman Society.[7]

A similar attitude toward health and illness was found in ancient Near Eastern cultures as well, where disease incurred personal blame. Throughout the Hebrew scriptures one finds the popular view enunciated that illness and disease are God's punishment for sin and wrongdoing. One does not pity the sick person but encourages that person to repent. This is the attitude of Bildad the Shuhite, one of Job's comforters, who warns Job that Yahweh acts justly; Job and his sons have sinned against him, but if he repents and remains upright in his behavior, Yahweh will prosper him (Job 8:1–10). Bildad's attitude transcends cultural boundaries; one finds it everywhere in the ancient world.[8]

The Christian perspective on sickness and health marked a dramatic departure from this view. The value that Christians accorded to suffering gave to the sick person a positive status. In the literature of the early monastic movement those who were ill suffered no stigma. They were neither held responsible for their illness nor ostracized as having committed a sin for which they were being punished. They were not responsible for the restoration of their own health but were regarded as deserving of compassion and assistance from those within the monastery.[9] Thus Gregory of Nazianzus writes of Basil: "[Basil] however it was, who took the lead in pressing upon those who were men, that they ought not to despise their fellow men, nor to dishonor Christ, the one head of all, by their inhuman treatment of them; but to use the misfortunes of others as an opportunity of firmly establishing their own lot, and to lend to God that mercy of which they stand in need at his hands."[10] Basil's attitude extended particularly to the treatment of lepers, whose disease classical society regarded as being loathsome and contagious.[11] In late pagan culture medical explanations of the disease and of contagion often led to the isolation and social ostracism of lepers. But Christian bishops never attributed the disease and the suffering it brought to God's punishment. Rather, they considered lepers to be victims of misfortune, whether it was the result of physical or social factors. Nor did the Cappadocians regard them as ritually impure, and in this regard they shared the views of classical society. Gregory of Nazianzus considered leprosy a sacred disease in what was apparently the first occasion that the appellation had been used of this disease. The reason has nothing to do with etiology, for Gregory accepted the view of medical writers that it was due to an excess of black bile. Rather, it was the terrible nature of leprosy that produced holiness in those who were physically afflicted, but in a special way that was associated with their Christ-like suffering. Although Lazarus is not specifically named as having suffered from leprosy in the narrative of Jesus's parable (Lk. 16:19–31), he came to be regarded as the biblical model of a sacred beggar. Lepers, banished from society and covered with loathsome sores, were seen to be set apart for God and to bear the image of Christ and hence more worthy of honor than the healthy. Susan Holman speaks of the "Christian resocialization of the leper" that the Cappadocians justified on theological grounds and for which medicine provided an appropriate metaphor in the image of reverse contagion.[12] The Incarnation of Christ had changed and elevated the human body, including that of lepers. Their bodies transmitted their holiness to those who had cared for them. In a striking metaphor, they were said to bring healing to their caregivers, those who suffered from the spiritual diseases, such as greed, that so often afflict the physically healthy.

Christian understanding of suffering as salutary, moreover, not only deprived the sick person of stigmatization but gave him or her a purpose with which to endure suffering.[13] The Christian life constituted a fellowship of suffering that united Christians with their Lord and with other believers. The purpose of suffering in the Christian life was to edify and to prepare one for eternal glory in heaven. The value of suffering for the Christian is a theme that is found in the fathers, mixed (especially in the Greek fathers) with an element of popular Stoicism that sometimes seems to glorify suffering but in fact attempts to persuade Christians who wish to avoid suffering that they should accept it for their soul's good.

Early Christianity was characterized by a strong emphasis on philanthropy that urged both individual and corporate care of those in need. Graeco-Roman paganism lacked any religious or philosophical basis for charity that encouraged a personal concern for those in physical need. Henry Sigerist writes: "The sick man, the cripple, the weakling are less worthwhile men and can only be reckoned as such in the view of society. Their worth is determined solely in terms of the possibility for bettering their condition. A lifetime of sickness was completely despised. Antiquity offers no evidence of any provision for the care of the crippled. A sick man must become well again in order to count again as a worthwhile person."[14] In this sense one can say that paganism had no understanding of the social implications of medicine, a fact that cannot be attributed merely to its lack of a philanthropic spirit or its failure to provide public or private institutions for the sick. Pagan culture discouraged all attempts to deal with the sick as a societal problem, in part, because it assumed that the sick were suffering deservedly; in part, because of a pessimism that regarded society as incapable of significant improvement; and, in part, because of a quietism that rejected the desirability of attempting real change in society. The resulting passivity accounts for the failure of state officials to undertake public relief during times of plague and reflects the ease with which ancient societies accepted suffering without undertaking efforts to ameliorate it. Underlying it as well was the belief that plague was retributive, a punishment by the gods on society for some failure of an individual or magistrate that could be removed only by their propitiation. Finally, Crislip observes that, "in contrast with the medical obsession that had so consumed members of the aristocracy [beginning in the second century after Christ], monastic leaders wrestled less with the interpretation of sickness within their own bodies than with the treatment of the sick within society, that is, the health care system."[15]

Unlike the classical world, Christianity rooted its attitude to philanthropy in theology. The impulse behind Christian philanthropy was the encouragement of a self-giving love of one's fellow human beings that reflected the love of God in the Incarnation of Christ and his death for the redemption of the world. Jesus's parable of the Good Samaritan furnished the pattern for the Christians' care of the sick: every stranger in need was a neighbor who bore the image of God and to whom the love of God ought to be demonstrated. The suffering that Christians had experienced under persecution prepared them to engage in a program of comfort, consolation, and encouragement, first to fellow Christians and then to others outside the community of faith. The new emphasis on compassion as a necessary Christian virtue lifted the treatment of the sick beyond the classical emphasis on medical professionalism. One can speak of a difference in outlook that was grounded in very different worldviews. While it seems Whiggish to attribute to early Christians the beginnings of the concept of social medicine, we see the seeds of it within the context of Christian medical philanthropy. During the first three centuries, it was impossible for Christians who were undergoing persecution to establish permanent institutions for the care of the sick. But they carried out an active program of palliative care through hundreds of churches in cities throughout the Roman Empire. Wherever a church was founded (and the church was an urban institution), it became a focal point for the care of the sick. Plagues had the effect of forcing Christians to undertake the care of the sick to those outside the church. Permanent charitable institutions sprang up within a generation or two after the end of the persecution of the Christians. By the end of the fourth century hospitals had been founded throughout the Eastern Empire, from which they were transplanted to the West. Andrew Crislip has argued that the creation of the monastic infirmary led to the emergence of "a new social role for the sick" in late antiquity.[16] In fact, the evidence suggests just the reverse: it was the existence of a new attitude toward the care of the sick in the early church that provided the ideology that undergirded the creation of both the monastic infirmary and the hospital, as well as the varied charitable institutions that sprang up throughout the East. Nearly three centuries of experience in providing parochial care for the sick allowed institutional care to develop quickly once persecution ended and monastic communities could provide the necessary institutional framework and manpower.

Does this concern for the sick entitle Christianity to be called, as Adolf Harnack thought it did, a "religion of healing"? The answer must be a negative one simply because early Christianity did not promise physical healing to its converts. The emphasis on *caring* more than on *curing* constituted the chief ministry of the early Christian community to the sick, although the boundary was always blurred, and there was much overlap. Christians sought, however, to fulfill the words of Jesus, "I was sick and you took care of me" (Mt. 25:36). The care of the sick was initially a duty incumbent especially upon deacons and deaconesses, although all Christians were expected to honor Jesus's injunction. Henry Sigerist was correct when he wrote that Christianity introduced the "most revolutionary and decisive change in the attitude of society toward the sick. . . . The social position of the sick man thus became fundamentally different from what it had been before. He assumed a preferential position which has been his ever since."[17] *This* was Christianity's novel contribution to health care. Pagan culture had no care of the sick organized on a charitable or community-wide basis, and there is no evidence to indicate that the Jewish community extended medical care beyond its own.[18] The Christian church offered its philanthropy not only to Christians but to others as well, as was evident during the plague of the third century in the urban centers of Alexandria, Carthage, and Rome. When it became possible for Christians to offer palliative nonprofessional and (where resources were available) professional medical care in the founding of the hospital and related institutions, they did so.

Sigerist overstates his case, however, when he writes that "Christianity came into the world as the religion of healing," promising "a restoration both spiritual and physical."[19] On the contrary, I have argued that the evidence overwhelmingly indicates that Christianity did *not* promise physical healing.[20] Nor should one speak of the care of the sick as a special "healing ministry" of the early church. It was an important part, but only a part, of the general philanthropic outreach of the church, which included caring for widows and orphans, aiding the poor, visiting those in prison, and extending hospitality to travelers. To the historian of medicine the contribution of Christianity to the care of the sick was fundamental and far reaching, particularly in its later institutionalization in the hospital and other more specialized institutions of health care. But in contrast to the great healing religions of the classical world, such as the cults of Asclepius, Isis, and Serapis, Christianity was not chiefly concerned with physical healing during the early centuries of its existence except as it was a component of its charitable ministry to those in need. The theme of "Christus medicus" (Jesus the Physician) was indeed a prominent one in the early church. But the phrase was used metaphorically to refer to Jesus as the healer of the afflictions of the soul, and only rarely of the diseases of the body. Even in the fifth century, when miraculous healing came to achieve greater (though still limited) prominence in Christianity, a prominence that it was to maintain for centuries, redemption—the salvation of the soul—remained paramount.

Vivian Nutton asserts that there existed no tension with medicine in paganism but rather a "striking collaboration between priest and doctor,"[21] while Christianity "introduced new tensions into the relationship between religion and medicine."[22] He adduces specifically the offer of miraculous healing as an alternative medicine and cites the familiar passage in James 5:13–18 as providing the basis for a sacramental rite. Even *had* Christianity held out the promise of supernatural healing to complement secular medicine, it would not necessarily have created tension. But apart from the passage in James there exists in the New Testament no blanket promise of miraculous healing that extends to all believers (Mark 16:18 is a non-authentic late interpolation that is not found in the earliest manuscripts). The passage in James, I have argued, speaks of healing from sin rather than physical healing, and it was so interpreted by the earliest commentators. It was too ambivalent to be pressed into service in support of a rite of sacramental healing of the body, and it was not so used in the first centuries of Christianity. It is equally mistaken to assert that early Christians encouraged the sick to seek miraculous healing for their diseases and, conversely, that Christians *always* urged them to seek the aid of physicians. In fact, one finds medicine and religion spoken of not only in a harmonious but even in a mutually supportive fashion in the writings of the church fathers. In part this arises from the linguistic ambiguity in the concept of salvation in Jewish and Christian thought by which the healing of the body served as an appropriate metaphor of the healing of the soul. Among the Cappadocians in the fourth century we find a particularly close cooperation between medicine and Christian philanthropy. Thus Gregory of Nyssa challenged those who avoided contact with lepers because of the potential for contagion by arguing that there are no *medical* grounds for supposing that the disease can be transmitted by touch. He writes that while plagues have an external cause in environmental factors leprosy cannot be passed from those afflicted with it to the healthy because "it is only in the interior that the illness develops, invading the blood by putrid humors which infect it and the infection does not leave the sick person."[23] Here he brings a medical understanding of leprosy into the service of urging Christians to overcome their instinctive reticence to engage in personal contact with those whom society considers untouchable, indeed repulsive.

The possibility of tension arises, however, in the two exceptions to the assumption that Christians would ordinarily seek the treatment of physicians or other forms of natural healing. Origen urged that exceptional Christians look to God for healing by prayer alone. We have instances of Christians who accepted his advice, most of them ascetics who would not receive medical treatment for themselves, while at the same time they ministered to the physical needs of the sick.[24] "To the ascetic," writes Susan Ashbrook Harvey, "illness was simply one more form of suffering to endure in the imitation of Christ."[25] As the body was brought under subjection in illness, so the soul grew through suffering. Such an ascetic was John of Ephesus (c. 507–89), who, while refusing medical treatment for his own illnesses, was concerned to treat other ascetics when they needed treatment. He relates how the ascetic Aaron, whose loins were eaten away by gangrene, had a leaden tube inserted so that he could pass water and lived for another eighteen years. There is no tension here. Even ascetics sometimes required medical treatment. Harvey suggests that there existed an "expedient alliance" between ascetics and doctors for pragmatic reasons, in order to allow God's servants like Aaron to continue in their service when workers in his kingdom were scarce.[26] Andrew Crislip argues, however, that ascetics recognized the value of medicine and were not averse to accepting medical treatment even for themselves when it was called for.[27] Whichever explanation is correct (and I believe that Crislip's is more likely), the tension is minimal. The issue is not whether the use of medicine is wrong for a Christian, since it is very difficult to find *any* Christians in the early church who held that position *on theological grounds*. The question is rather whether there were instances in which one should refuse medicine as a gift of God for a greater good, the good of the soul.

The second exception is the belief among some early Christians that when illness came as God's chastisement for sin, medicine should be refused. The connection between sin and suffering was not limited to Christians; it was age-old. Basil, in recommending in the *Long Rule* (55) that monks abstain from medicine if they believe that God is chastising them, urges them to use their suffering as a means of spiritual growth. His advice represents a characteristically and uniquely Christian approach: that suffering is sent by God as a means of grace for the spiritual benefit of the sufferer. It invites introspection. Is God speaking through one's suffering? If so, what is to be learned from it? Even if the suffering was not the result of sin, self-examination could result in spiritual illumination or purification. The decision whether to accept or reject medicine was a personal one. It was not prescribed but remained the decision of the individual. If it appears that Christian writers exalt suffering, it is because they endue it with meaning and urge that the believer seek to understand whether God is speaking through their experience with illness. But they did not posit a simple correlation between sin and suffering; rather, illness was a means by which one might avoid sin. Meredith writes: "In the case of the saint, there is no obvious link between sin and bodily affliction. The assumption of the saint's holiness precludes understanding his illnesses as resulting from sin. For the holy man, illness is not an outward sign of inward sin; it is not an external articulation of moral failings. Affliction does not reveal hidden sinfulness, rather, it is yet another manifestation of the holy man's saintliness."[28]

In a well-known statement, Ludwig Edelstein contrasts the high value that the classical world placed on health with what he terms "the Christian or Romantic glorification of disease."[29] Vivian Nutton echoed Edelstein's contrast between the alleged Christian attitude to disease and that of the classical world: "[B]ut it is only with Judaism, and, still more, with Christianity, that one finds passive acceptance of disease and suffering enjoined upon the true believer because they are part of God's judgement."[30] Nutton cites Cyprian, who urges his congregants to accept the plague joyfully because by it the wicked will more speedily be sent to hell and the righteous will more quickly attain eternal life.[31] But Nutton quotes only a portion of Cyprian's words. Cyprian continues:

> How suitable, how necessary it is that the plague and pestilence, which seems horrible and deadly, searches out the justice of each and every one and examines the minds of the human race; whether the well care for the sick, whether relatives dutifully love their kinsmen as they should, whether masters show compassion for their ailing slaves, whether physicians do not desert the afflicted. . . . Although this mortality has contributed nothing else, it has especially accomplished for Christians and servants of God, that we have begun gladly to seek martyrdom while we are learning not to fear death. These are trying exercises for us, not deaths; they give to the mind the glory of fortitude; by contempt of death they prepare for the crown.[32]

Cyprian's statement, when taken out of context, appears smug and judgmental. In fact, he cites the plague as the bearer of death (and so of final judgment) that in the end awaits all men. But he particularizes it by asking questions that will search the consciences of his readers regarding how they have responded to the call to compassionate care of those in need. The sermon urges the Christian community, in a time of intense persecution in which Cyprian himself lost his life, to undertake the systematic care of the sick throughout the entire city, even if it involves aiding their persecutors. He goes on to say: "There is nothing remarkable in cherishing merely our own people with the due attentions of love, but that one might become perfect who should do something more than heathen men or publicans, one who, overcoming evil with good, and practicing a merciful kindness like that of God, should love his enemies as well. . . . Thus the good was done to all men, not merely to the household of faith."[33]

Nutton also cites Tertullian, who sees famine and plague as God's means of limiting excessive prosperity and overpopulation.[34] Tertullian's assertion should probably be taken as nothing more than a reflection of the Christian tendency to look for evidence of God's providence in troubling circumstances. In fact, early Christian writers like Cyprian and Tertullian, while maintaining a very different understanding of the role of disease and suffering than that of classical writers, did not encourage a merely passive attitude in the face of suffering; nor did they urge Christians to seek suffering for themselves. Andrew Crislip cites the case of Simeon Styletes, who deliberately sought to mortify the flesh by wrapping a rope tightly around his body until, after a year, the lacerated flesh produced an offensive stench.[35] The abbot reprimanded him and asked him to leave the monastery. "For Basil," who insisted that the desire for sickness to purge the body must be moderated, "the pursuit of suffering is no better than the pursuit of luxury; both entail an excessive focus on the body."[36]

Cyprian and Tertullian, in responding to the plague of the third century, formulated a view of the human condition in which suffering assumed a positive role that it had previously lacked in the ancient world.[37] "The epidemics," writes Stark, "swamped the explanatory and comforting capacities of paganism and Hellenic philosophies. In contrast, Christianity offered a much more satisfactory account of why these terrible times had fallen upon humanity, and it projected a hopeful, even enthusiastic, portrait of the future."[38] Suffering became an essential component of the process of Christian sanctification that, when humbly accepted, possessed great spiritual value. Rather than bringing shame and disapproval, disease and sickness gave to the sufferer a favored status that invited sympathy and compassionate care. The two attitudes can be seen juxtaposed in a festal letter of Dionysius of Alexandria, written during the plague, which is preserved by Eusebius. In it Dionysius contrasts the joy of the Christians who cared for the sick and, in so doing, often contracted the plague and died, with the fear of their pagan neighbors who abandoned the sick and fled the city.[39] Even if one makes allowance for the triumphalist tone of the letter, it is clear that widely different attitudes prevailed among the population of plague-ridden Alexandria: "The heathen behaved in the very opposite way. At the first onset of the disease, they pushed the sufferers away and fled from their dearest, throwing them into the roads before they were dead, and treating unburied corpses as dirt, hoping thereby to avert the spread and contagion of the fatal disease; but do what they might, they found it difficult to escape" (7:22). Had this description of plague psychology not echoed earlier descriptions by classical authors, one might dismiss it as mere sectarian propaganda. In fact, the response that Dionysius attributed to many Alexandrians was little different from that which Thucydides described during the plague of Athens. It constituted the traditional attitude of those brought up within a classical culture in which neither philosophy nor religion encouraged a compassionate response to human suffering.

If indeed a potential for tension existed, it is to be found in the dualism that accompanied the introduction of asceticism into Eastern Christianity in the third century. But the earliest form of asceticism, particularly in Syrian Christianity, was not based on a body-soul distinction but rather on the belief that denying the body meant dying to the world in order to live in God's kingdom, the present world forming a continuum with the world to come.[40] Something very like dualism informs the invitation to physical suffering that was glorified by some ascetics such as Macarius; yet, as Peter Brown has pointed out, even in the midst of what appears to us to be extreme Christian asceticism there remained the belief that the body was destined for resurrection, which spurred monks to ascetic discipline in order to purify a body that would someday be transformed.[41] Moreover, the same call to asceticism that led monks to mortify their bodies also led them to provide medical care for the sick, especially the poor. But even ordinary ascetic practices within monasteries were suspended when circumstances demanded it. When monks were ill, they were granted additional food and drink, even meat and wine that were ordinarily prohibited. While some fellow monks complained at the release of the sick from restrictions of diet, those who were ill would be given special privileges as a matter of course in order to assist in their recovery.[42] And in the introspection that led monks to ask whether in their physical suffering they were being chastised by God for personal sins, there was no denial of the value of medicine, only the question of whether it was expedient for ascetics to employ it to relieve the discomfort of the body. One cannot speak of an ambivalent attitude of ascetics to medicine and physicians, only of a desire to renounce the good things of ordinary life for something better, a stronger dependence on God and his grace. The benefits of medicine, like food and drink, clothing and hygiene, were never in doubt. With rare exceptions there was no dualism in Christianity that held matter to be evil. Even while contemplating whether to refuse medicine or physicians, few ascetics ever doubted that they were acceptable if used legitimately. It is true that one finds much complaining about doctors in the hagiographical Christian literature of late antiquity, but it is a literary topos that owes a good deal to the denigration of physicians that is so often found earlier in classical literature.[43] It served the additional function of elevating the status of religious healers who could be shown to be successful in cases where doctors had failed.

Ascetic movements and the rise of monasticism had the potential for curtailing the positive role that medicine played in Christians' approach to healing. Far from denigrating secular medicine, however, monastics incorporated it into their religious vocation, a fact that is due in large part to the church's acceptance of the role of physicians as early as the second century.[44] Medicine was so much a universal component of the larger Graeco-Roman culture that Christians expressed little resistance to its appropriation, although they discussed its theological implications and debated whether there were special instances when one might forgo it, a feature that one would not find in pagan sources. So thoroughly did the Christian community assimilate it that when monks created the earliest charitable healing institutions in the fourth century, they staffed them, when financial resources and personnel became available, with physicians rather than religious healers. In so doing they joined Christian charity to the healing art, thereby adapting a motif in which, in the words of Macaulay, "the great Physician of the soul did not disdain to be also the physician of the body."[45] The result was a novel concept of healing that went beyond anything that the classical world had to offer: institutional health care administered in a spirit of compassion by those whose desire to serve God summoned them to a life of active beneficence.

ABBREVIATIONS

Amundsen	Darrel W. Amundsen, *Medicine, Society, and Faith in the Ancient and Medieval Worlds* (Baltimore: Johns Hopkins University Press, 1996)
ANF	A. Roberts and J. Donaldson, eds., *The Ante-Nicene Fathers,* 10 vols. (1885–96; repr., Grand Rapids, Mich.: Eerdmans, 1956)
ANRW	*Aufstieg und Niedergang der Römischen Welt*
BHM	*Bulletin of the History of Medicine*
CBQ	*Catholic Biblical Quarterly*
Crislip	Andrew T. Crislip, *From Monastery to Hospital: Christian Monasticism and the Transformation of Health Care in Late Antiquity* (Ann Arbor: University of Michigan Press, 2005)
DCA	William Smith and Samuel Cheetham, *A Dictionary of Christian Antiquities* (London: John Murray, 1875–80)
Edelstein	Ludwig Edelstein, *Ancient Medicine: Selected Papers of Ludwig Edelstein,* ed. Owsei Temkin and C. Lilian Temkin (Baltimore: Johns Hopkins Press, 1967)
ERE	James Hastings, *Encyclopedia of Religion and Ethics* (New York: Charles Scribner's Sons, 1908–26)
Finn	Richard Finn, OP, *Almsgiving in the Later Roman Empire: Christian Promotion and Practice (313–450)* (Oxford: Oxford University Press, 2006)
Grmek	Mirko D. Grmek, *Diseases in the Ancient Greek World,* trans. Mireille Muellner and Leonard Muellner (Baltimore: Johns Hopkins University Press, 1989)
JHM	*Journal of the History of Medicine and Allied Sciences*
Kühn	Karl Gottlob Kühn, ed., *Glaudii Galeni Opera Omnia,* 22 vols. (1821–33; repr., Hildesheim: Georg Olms, 1965)

Merideth	Anne Elizabeth Merideth, "Illness and Healing in the Early Christian East" (Ph.D. diss., Princeton University, 1999)
Stark	Rodney Stark, *The Rise of Christianity: A Sociologist Reconsiders History* (Princeton, N.J.: Princeton University Press, 1996)
SJT	*Scottish Journal of Theology*
TAPA	*Transactions and Proceedings of the American Philological Association*
TDNT	Gerhard Kittel and Gerhard Friedrich, eds., *Theological Dictionary of the New Testament,* trans. Geoffrey W. Bromiley (Grand Rapids, Mich.: Eerdmans, 1964–76)
Warfield	Benjamin B. Warfield, *Counterfeit Miracles* (1918; repr., London: Banner of Truth, 1972)

NOTES

Chapter 1 • Methods and Approaches

1. See, e.g., B. Palmer, ed., *Medicine and the Bible*, and K. Seybold and U. B. Mueller, *Sickness and Healing*, trans. D. W. Stott.

2. The limitations of our evidence are a problem that is not confined to sources for early Christian attitudes to medicine alone. See the remarks of Vivian Nutton on the fragmentary nature of our surviving sources for Greek medicine (*Ancient Medicine* 1–17).

3. J.-J. Von Allmen, ed., *A Companion to the Bible* 402, s.v. "Sickness" by H. Roux. See D. T. Reff, *Plagues, Priests, and Demons: Sacred Narratives and the Rise of Christianity in the Old World and the New* 20, regarding the role that epidemics played for Christian historians such as Eusebius and Gregory of Tours in illustrating God's power and providence.

4. See D. B. Wilson, "The Historiography of Science and Religion," and C. A. Russell, "The Conflict of Science and Religion," in *The History of Science and Religion in the Western Tradition: An Encyclopedia*, ed. G. Ferngren, 3–11 and 12–16.

5. D. C. Lindberg and R. L. Numbers, eds., *God and Nature: Historical Essays on the Encounter between Christianity and Science* 10.

6. J. H. Brooke, *Science and Religion: Some Historical Perspectives* 321.

7. D. Lindberg, "Science and the Early Church," in Lindberg and Numbers, *God and Nature* 19–48. See also his "Early Christian Attitudes toward Nature," in Ferngren, *The History of Science and Religion* 243–7.

8. Cf. Georg Luck's definition: "Miracles can be defined as extraordinary events that are witnessed by people, but that cannot be explained in terms of human power or the laws of nature. They are therefore frequently attributed to the intervention of a supernatural being" (G. Luck, *Arcana Mundi: Magic and the Occult in the Greek and Roman Worlds* 135 in the context of Luck's discussion of the problems inherent in defining miracles, 135–40).

9. For a detailed examination of the varieties and definitions of miracles, see H. Remus, *Pagan-Christian Conflict over Miracle in the Second Century*.

10. For a comprehensive discussion, see Luck, *Arcana Mundi* 3–60; Remus, *Pagan-Christian Conflict* 52–72, for a discussion of the language of magic in the classical world and the attempts by both pagans and Christians to demarcate it from miracle; and, for its definition, F. Graf, *Magic in the Ancient World*, trans. F. Philip, 8–19.

11. See J. G. Gager, *Curse Tablets and Binding Spells from the Ancient World* 24–25. Howard

Clark Kee, while recognizing that medicine, miracle, and magic are not separate and unrelated traditions, treats them as if they were (*Medicine, Miracle, and Magic in New Testament Times* 2–3). See the review by D. W. Amundsen in *BHM* 63 (1989): 140–4. Following E. E. Evans-Pritchard, anthropologists have tended in the past two generations to assume that no clear-cut distinction exists between magic and religion (D. E. Aune, "Magic in Early Christianity," in *ANRW* II. 23, 2 [1981]: 1510–6). On the distinction between magic and miracle, see Aune, "Magic in Early Christianity," 1521–2. On the differences between magic and religion in prayer in Graeco-Roman culture, see F. Graf, "Prayer in Magic and Religious Ritual," in *Magika Hiera: Ancient Greek Magic and Religion* 188–9, 196–7.

12. So D. B. Martin, *The Corinthian Body* 280 n. 61.

13. P. J. van der Eijk, *Medicine and Philosophy in Classical Antiquity: Doctors and Philosophers on Nature, Soul, Health and Disease* 9 n. 17.

14. Remus, *Pagan-Christian Conflict* 37–39. Remus cites several examples, of which the best known is the polemic of the writer of the Hippocratic treatise *The Sacred Disease.*

15. Ibid., 39–40.

16. G. Clarke, *Women in Late Antiquity* 66.

17. A current view depicts Jesus as a wandering charismatic whose followers were popular healers and itinerant Cynic-like evangelists (see G. Theissen, *Sociology of Early Palestinian Christianity,* trans. J. Bowden; J. D. Crossan, *The Historical Jesus: The Life of a Mediterranean Jewish Peasant;* and M. Borg, *Jesus: A New Vision*). For a balanced discussion of the assumptions that underlie any reading of the Gospels, see N. T. Wright, *The New Testament and the People of God,* vol. 1 of *Christian Origins and the Question of God* 3–120.

18. See A. N. Sherwin-White, *Roman Society and Roman Law in the New Testament* 186–93.

19. See G. E. Ladd, *The New Testament and Criticism* 171–214.

20. On the quest of the historical Jesus in its various manifestations and the literary and historical problems that present themselves in any attempt to understand and interpret Jesus, see N. T. Wright, *Jesus and the Victory of God,* vol. 2 of *Christian Origins and the Question of God* 3–144.

21. See, e.g., B. J. Malina, *The New Testament World: Insights from Cultural Anthropology* and *Christian Origins and Cultural Anthropology.* On the ideological premises that underlie contemporary anthropological approaches (the structural-functional method and the phenomenological method in comparative religion), see Aune, "Magic in Early Christianity" 1509, and Stark 79 and 166–7.

22. J. J. Pilch, *Healing in the New Testament: Insights from Medical and Mediterranean Anthropology*; H. Avalos, *Health Care and the Rise of Christianity.*

23. See G. E. R. Lloyd, *In the Grip of Disease: Studies in the Greek Imagination* 1–13. For a discussion of Brown's indebtedness to social anthropology see P. A. Hayward, "Demystifying the Role of Sanctity in Western Christendom," in *The Cult of Saints in Late Antiquity and the Middle Ages,* ed. J. Howard-Johnston and P. A. Hayward, 115–31; É. Rebillard, "La 'Conversion' de l'Empire romain selon Peter Brown (note critique)," *Annales HSS* 54:4 (1999): 813–23; and P. Brown's own description of the successive influences on his work in "The Saint as Exemplar in Late Antiquity," *Representations* 2 (1983): 1–25 passim.

24. See my review of Pilch's *Healing in the New Testament, BHM* 78 (2004): 468–9.

25. P. Brown, *Poverty and Leadership in the Later Roman Empire* 107.

26. Stark 209–10. Cf. Han Drijvers, who speaks of "the interrelation between sociological and ideolgical elements in society" (quoted by Brown, "The Saint as Exemplar in Late Antiquity" 14).

27. See D. Praet, "Explaining the Christianization of the Roman Empire: Older Theories and Recent Developments," *Sacris Erudiri: Jaarboek voor Godsdienstwetenschappen* 33 (1992–93): 7–119; L. M. White, "Adolf Harnack and the Expansion of Early Christianity," *Second Century* 5 (1985/86): 97–127; and Reff, *Plagues, Priests, and Demons* 8–34, to whose discussion I am indebted for much of what follows.

28. See, for example, J. Perkins, *The Suffering Self: Pain and Narrative Representation in the Early Christian Era.*

29. See I. Strenski, "Religon, Power, and Final Foucault," *Journal of the American Academy of Religion* 66, no. 2 (1998): 345–67.

30. Ibid., 345–52.

31. So A. Cameron, *Christianity and the Rhetoric of Empire: The Development of Christian Discourse* 2–7, and T. E. Klutz, "The Rhetoric of Science in *The Rise of Christianity*: A Response to Rodney Stark's Sociological Account of Christianization," *Journal of Early Christian Studies* 6 (1998): 180–4. The last two terms are Klutz's (184).

32. Reff, *Plagues, Priests, and Demons* 36–7.

33. Cf. the remarks of Robert A. Markus, who distinguishes between boundaries drawn by the practitioners of modern disciplines ("twentieth-century historians, anthropologists, theologians") and those drawn by the people of late antiquity whom they study. Both are legitimate and necessary but should not be confused in defining concepts employed in the historical reconstruction of the past (*The End of Ancient Christianity* 8–9).

34. One cannot make a clear-cut distinction between Palestinian and Hellenistic Judaism given the extent to which Hellenistic culture had penetrated Palestine (Aune, "Magic in Early Christianity" 1519 and n. 46 for an extensive bibliography).

35. Three types of religion existed in the classical world: state cults, mystery cults, and philosophical sects. Christianity was something of an anomaly, being a religion that was not related to nationality (ibid., 1519–20).

36. Cf. Rebecca Flemming, who writes that "across the Roman Empire a Hellenistic medical *koinē*, a communal conceptual and practical holding, was established and developed, subject to regional variation but recognizably of the same stock" (*Medicine and the Making of Roman Women: Gender, Nature, and Authority from Celsus to Galen* 51).

37. Cf. the remarks of Martin, *The Corinthian Body* xiii–xiv.

38. See B. D. Ehrman, *Lost Christianities: The Battles for Scripture and the Faiths We Never Knew.* Unlike Ehrman, however, I do not consider early Christianity to be merely a heuristic construction. See the remarks of R. Stark, "E Contrario," *Journal of Early Christian Studies* 6 (1998): 260–2.

39. Cf. David Lindberg's definition of "the Christian position" as "the 'center of gravity' of a distribution of Christian opinion, for great variety existed" (quoted by Amundsen 5).

40. W. H. C. Frend, *The Rise of Christianity* 230.

41. R. Lane Fox, *Pagans and Christians* 31.

42. See M. Kahlos, *Vettius Agorius Praetextatus: A Senatorial Life in Between* 4–8, and Markus, *The End of Ancient Christianity* 28.

43. E.g., R. MacMullen, *Christianizing the Roman Empire (A.D. 100–400)* 74–85.

44. E.g., R. MacMullen, *Christianity and Paganism in the Fourth to Eighth Centuries* 74–102 ("Superstition") and 103–49 ("Assimilation"). Markus notes the uncritical use of terms like "pagan survivals" and "christianisation," which cloak complex cultural issues (*The End of Ancient Christianity*) 9–16.

45. D. Lindberg, "Medieval Science and Religion," in Ferngren, *The History of Science and Religion* 266.

Chapter 2 • The Christian Reception of Greek Medicine

1. "*Disease* refers to a malfunctioning of biological and/or psychological processes, while the term *illness* refers to the psychosocial experience and meaning of perceived disease" (A. Kleinman, *Patients and Healers in the Context of Culture: An Exploration of the Borderland between Anthropology, Medicine, and Psychiatry* 72). Cf. Lloyd, *In the Grip of Disease* 1–2.

2. See B. J. Good, *Medicine, Rationality, and Experience: An Anthropological Perspective* 101–15, on the incommensurability of medical systems, with special reference to the humoral system.

3. Merideth 23–5.

4. The ontological theory was anticipated by the Methodist sect. Some Methodists maintained that disease was caused by invisible "seeds" (*animalcula*) and transmitted from person to person through contact (V. Nutton, "The Seeds of Disease: An Explanation of Contagion and Infection from the Greeks to the Renaissance," *Medical History* 27 (1983): 1–34; reprinted in *From Democedes to Harvey: Studies in the History of Medicine* XI 1–34).

5. D. W. Amundsen and G. B. Ferngren, "The Perception of Disease and Disease Causality in the New Testament," in *ANRW* II. 37, 3 (1996): 2935–6.

6. On theories of disease in the Hippocratic Corpus, see Nutton, *Ancient Medicine* 72–86; on causation, see R. J. Hankinson, "Galen's Theory of Causation," in *ANRW* II. 37, 2 (1994): 1757–74.

7. V. Nutton, "Humoralism," in *Companion Encyclopedia of the History of Medicine*, ed. W. F. Bynum and R. Porter, 1:281–91; O. Temkin, "The Scientific Approach to Disease: Specific Entity and Individual Sickness," in *The Double Face of Janus and Other Essays in the History of Medicine* 442–8; Hankinson, "Galen's Theory of Causation" 1762–3. Galen claimed, however, that in his day Hippocratism was a minority view.

8. Grmek 7–8.

9. C. J. F. Poole and A. J. Holladay, "Thucydides and the Plague of Athens," *Classical Quarterly* 29 (1979): 282–300.

10. Grmek 13.

11. This is the conclusion reached by Poole and Holladay after their careful examination of previous attempts to identify the plague of Athens ("Thucydides and the Plague of Athens" 286–95); regarding subsequent attempts, see A. J. Holladay, "New Developments in the Problem of the Athenian Plague," *Classical Quarterly* 38 [1988]: 247–50. Thucydides describes the symptoms in his *History* 2.47–54. Parry argued that he does not employ a technical medical vocabulary but uses words that are found in both medical and nonmedical writers (A. Parry, "The Language of Thucydides' Description of the Plague," *Bulletin of the Institute of Classical Studies* 16 [1969]: 106–18). Thucydides' narrative exhibits a dramatic and moral tone,

but the extent of its dependence on Hippocratic treatises remains a matter of debate. A large literature exists on the subject; see particularly S. Hornblower, *A Commentary on Thucydides* I: 316–18; T. E. Morgan, "Plague or Poetry? Thucydides on the Epidemic at Athens," *TAPA* 124 [1994]: 197–209; S. Swain, "Man and Medicine in Thucydides," *Arethusa* 27 (1994): 303–28; and R. Mitchell-Boyask, *Plague and the Athenian Imagination: Drama, History and the Cult of Asclepius* 41–3. DNA examination of skeletal material from a mass burial pit discovered in 1994–95 in the Athenian Kerameikos that dates roughly from the time of the plague of Athens (c. 430 B.C.) has not so far made possible a certain identification of the disease: see E. Baziotopoulou-Valavani, "A Mass Burial from the Cemetery of Kerameikos," in *Excavating Classical Culture: Recent Archaeological Discoveries in Greece* 187–201, and M. Papagrigorakis, C. Yapijakis, S. Phillippos and E. Baziotapoulou-Valavani, "DNA Examination of Ancient Dental Pulp Incriminates Typhoid Fever as Probable Cause of the Plague of Athens," *International Journal of Infectious Diseases* 10 (2006): 206–14.

12. Grmek 6–14; Nutton, *Ancient Medicine* 31–3; K.-H. Leven, " 'At Times These Ancient Facts Seem to Lie before Me like a Patient on a Hospital Bed'—Retrospective Diagnosis and Ancient Medical History," in *Magic and Rationality in Ancient Near Eastern and Graeco-Roman Medicine,* ed. H. F. J. Horstmanshoff and M. Stol, 369–86. John Riddle is more hopeful; see "Research Procedures in Evaluating Medieval Medicine," in *The Medieval Hospital and Medical Practice,* ed. Barbara S. Bowers, 3–17.

13. Nutton, *Ancient Medicine* 28–9.

14. Ibid., 31–3; R. Sallares, *Malaria and Rome: A History of Malaria in Ancient Italy* 9–11.

15. Nutton, *Ancient Medicine* 22–3.

16. Poole and Holladay, "Thucydides and the Plague of Athens" 294–5.

17. D. J. Ladouceur, "The Death of Herod the Great," *Classical Philology* 76 (1981): 25–34.

18. K.-H. Leven, "Athumia and Philanthropia: Social Reactions to Plagues in Late Antiquity and Early Byzantine Society," in *Ancient Medicine in Its Socio-Cultural Context,* ed. P. J. van der Eijk et al., 2:393–407.

19. E. Lieber, "Old Testament 'Leprosy,' Contagion and Sin," in *Contagion: Perspectives from Pre-Modern Societies,* ed. L. I. Conrad and D. Wujastyk, 99–136.

20. Grmek 152–76, 198–204; K. F. Kiple, ed., *The Cambridge World History of Human Disease* 834–9; K. Manchester, "Leprosy: The Origin and Development of the Disease in Antiquity," in *Maladie et maladies: Histoire et conceptualisation,* ed. D. Gourevitch, 31–49; S. R. Holman, "Healing the Social Leper in Gregory of Nyssa's and Gregory of Nazianzus's 'περὶ φιλοπτωχίας,' " *Harvard Theological Review* 92 (1999): 287–93.

21. Holman, "Healing the Social Leper" 292.

22. Ibid., 285–6.

23. Grmek 133–51; Kiple, *The Cambridge World History of Human Disease* 834–9, 756–63 (gonorrhea), 1053–5.

24. Nutton, *Ancient Medicine* 26–7.

25. V. Nutton, "Did the Greeks Have a Word for It? Contagion and Contagion Theory in Classical Antiquity," in Conrad and Wujastyk, *Contagion* 137–62.

26. Ibid., 151.

27. Ibid., 157.

28. *Iliad* 11.514 ff.; cf. *Odyssey* 17.382–6.

29. R. Brock, "Sickness in the Body Politic: Medical Imagery in the Greek Polis," in *Death and Disease in the Ancient City*, ed. V. M. Hope and E. Marshall, 24–34; cf. C. Schubert, "Menschenbild und Normwandel in der klassischen Zeit," in *Médecine et morale dans l'antiquité*, ed. H. Flashar and J. Jouanna, 121–43.

30. Cf. J. Jouanna, *Hippocrates*, trans. M. D. DeBevoise, 181–209. G. E. R. Lloyd makes important qualifications in "The Invention of Nature," in *Methods and Problems in Greek Science* 420–4.

31. "Quite fantastical" is Lloyd's description (G. E. R. Lloyd, *Science and Morality in Greco-Roman Antiquity* 12–13). Philip van der Eijk demonstrates that the author does not altogether reject either divine intervention or religious healing ("The 'Theology'of the Hippocratic Treatise, *On the Sacred Disease*," in *Medicine and Philosophy in Classical Antiquity: Doctors and Philosophers on Nature, Soul, Health and Disease* 67–8).

32. On the training of physicians and on medical schools, see V. Nutton, "Museums and Medical Schools in Classical Antiquity," *History of Education* 4 (1975): 3–15.

33. On the Roman appropriation of Greek medicine, see V. Nutton, "Roman Medicine: Tradition, Confrontation, Assimilation," in *ANRW* II. 37, 1 (1993): 49–78.

34. D. Gourevitch, *Le Triangle hippocratique dans le monde gréco-romain: Le Malade, sa maladie et son médecin* 289–322.

35. Pliny, *Natural History* 29.7 (who quotes Cato); Pliny the Younger, *Letters* 2:20.4–5 (on soothsayers).

36. On the medical sects generally, see D. Gourevitch, "The Paths of Knowledge: Medicine in the Roman World," in *Western Medical Thought from Antiquity to the Middle Ages*, ed. M. D. Grmek, 104–38.

37. Celsus, *De medicina*, Prooemium 13, 25–26.

38. Ibid., 27–43.

39. Nutton, *Ancient Medicine* 187–201.

40. John Riddle believes that ancient drugs were more often efficacious than is usually thought ("Research Procedures in Evaluating Medieval Medicine" 8–14).

41. *Aphorisms* 7.87.

42. On the naturalism of the Hippocratic Corpus, see G. E. R. Lloyd, *Magic, Reason and Experience: Studies in the Origin and Development of Greek Science* 10–37. Elsewhere Lloyd observes that there existed everywhere in Greece and at all times the *implicit assumption* that natural phenomena are regular. But the development of an *explicit concept* was forged in debates until there came to exist "almost as many theories of nature as of . . . earthquakes" ("The Invention of Nature" 419 and 432).

43. Lloyd, *Science and Morality* 12.

44. G. E. R. Lloyd, as quoted by J. M. Riddle, "Folk Tradition and Folk Medicine: Recognition of Drugs in Classical Antiquity," in *Folklore and Folk Medicines*, ed. J. Scarborough, 40.

45. Riddle, "Folk Tradition and Folk Medicine" 40–2; J. Scarborough, "Adaptation of Folk Medicines in the Formal Materia Medica of Classical Antiquity," in *Folklore and Folk Medicines* 21–32; Lloyd, *Magic, Reason and Experience* 42–5; Luck, *Arcana Mundi* 38–9, 71–3.

46. G. E. R. Lloyd, *Science, Folklore and Ideology: Studies in the Life Sciences in Ancient Greece* 202–3.

47. Ibid., 119–49 (on Theophrastus, the Hippocratics, and Pliny).

48. Riddle, "Folk Tradition and Folk Medicine" 40–1.

49. A. Rousselle, "From Sanctuary to Miracle-Worker: Healing in Fourth-Century Gaul," in *Ritual, Religion, and the Sacred: Selections from the Annales, Économies, Sociétés, Civilisations,* vol. 7, ed. R. Forster and O. Ranum, trans. E. Forster and P. M. Ranum, 101–9.

50. Rousselle overstates the case when she asserts that "Christianity continued to see all illness, madness, and all the tribulations that the pagans attributed to maleficent powers as signs that the demon had gained power over a creature" (ibid., 112). The culture (or more likely some within the culture) ascribed *some* diseases and conditions to demons, however, as can be seen in the reported healings of Martin of Tours (d. 397) (ibid., 110–24).

51. L. Edelstein, "The Hippocratic Physician," in Edelstein 87–110; H. F. J. Horstmanshoff, "The Ancient Physician: Craftsman or Scientist?" *JHM* 45 (1990): 176–97.

52. Suetonius, Julius 42.1; Flemming, *Medicine and the Making of Roman Women* 44.

53. Nutton, "Roman Medicine" 62, 69–70, 74–6.

54. V. Nutton, "Healers in the Medical Market Place: Towards a Social History of Graeco-Roman Medicine," in *Medicine and Society,* ed. A. Wear, 39. Of the forty physicians in Gaul who are known from inscriptions in Roman imperial times, most have Greek names or surnames. Greek was the language of physicians in the West (Rousselle, "From Sanctuary to Miracle-Worker" 100).

55. Rousselle, "From Sanctuary to Miracle-Worker" 106–7. According to Vivian Nutton, the epigraphic evidence indicates that 80 percent of physicians in the Western Empire were either slaves or freedmen in the first century. The percentage dropped to 50 percent in the second century and to 25 percent in the third (cited by Flemming, *Medicine and the Making of Roman Women* 51).

56. Lloyd, *Magic, Reason and Experience* 37–40. On competition for patients and polemics against medical opponents, see ibid., 86–98.

57. Lloyd, *In the Grip of Disease* 149.

58. See, e.g., Plato, *Republic* 406A–407D; D. W. Amundsen, "The Physician's Obligation to Prolong Life: A Medical Duty without Classical Roots," *Hastings Center Report* 8, no. 4 (1978): 23–30.

59. On the importance of dreams in diagnosis (especially in medical writers), see Lloyd, *Magic, Reason and Experience* 43. For one of many striking examples, that of the orator Libanius, see Gage, *Curse Tablets and Binding Spells* 121 and 124 n. 23. Aelius Aristides was often helped by dreams to find therapies for his many illnesses, some conventional and some most unconventional.

60. Lloyd, *In The Grip of Disease* 238–9. In a private communication John Riddle suggests that there was advance in medicine during this period but that it was largely attributable to improved water supplies and nutrition, not to formal medical delivery systems, although the impact of Hellenism and rationality would have eliminated some of the more unsuccessful therapies from primary medical services.

61. Seybold and Mueller, *Sickness and Healing* 16–96; N. Allan, "The Physician in Ancient Israel: His Status and Function," *Medical History* 45 (2001): 377–94; H. Avalos, *Illness and Health Care in the Ancient Near East: The Role of the Temple in Greece, Mesopotamia, and Israel* 233–99; M. Dörnemann, *Krankheit und Heilung in der Theologie der frühen Kirchenväter* 13–22.

62. H. D. Betz, ed., *The Greek Magical Papyri in Translation including the Demotic Spells* xlv and lii–liii n. 47, for bibliography; J. Goldin, "The Magic of Magic and Superstition," in *Aspects of Religious Propaganda in Judaism and Early Christianity*, ed. E. S. Fiorenza, 115–47; P. J. Achtemeier, "Jesus and the Disciples as Miracle Workers in the Apocryphal New Testament," in Fiorenza, *Aspects of Religious Propaganda* 152–6.

63. S. T. Newmyer, "Talmudic Medicine and Greco-Roman Science: Crosscurrents and Resistance," in *ANRW* II. 37, 3 (1996): 2904; S. S. Kottek, "Hygiene and Healing among the Jews in the Post-Biblical Period: A Partial Reconstruction," in *ANRW* II. 37, 3 (1996): 2843–65; F. Rosner, "Jewish Medicine in the Talmudic Period," in *ANRW* II. 37, 3 (1996): 2866–94. On the relative ease with which Jews of the diaspora became hellenized, see Stark 60–1.

64. Wisdom 38:1–15.

65. S. J. Noorda, "Illness and Sin, Forgiving and Healing: The Connection of Medical Treatment and Religious Beliefs in Sira 38, 1–5," in *Studies in Hellenistic Religions*, ed. M. J. Vermaseren, 215–24.

66. Amundsen and Ferngren, "Perception of Disease" 2952; cf. L. P. Hogan, *Healing in the Second Temple Period* 168–207, and Allan, "The Physician in Ancient Israel" 393–4. On Josephus, see S. Kottek, *Medicine and Hygiene in the Works of Flavius Josephus.*

67. See, e.g., J. M. Baumgarten, "The 4Q Zadokite Fragments on Skin Disease," *Journal of Jewish Studies* 41 (1990): 153–65, who identifies a physiological description of the flow of the blood in the arteries. Another text, written on leather, that was found among the Dead Sea Scrolls, 4Q Therapeia, was described by James Charlesworth as "cryptic notes on the medical rounds of a Jewish physician"; Charlesworth dates the text to A.D. 26–60 (J. H. Charlesworth, *The Discovery of a Dead Sea Scroll (4Q Therapeia): Its Importance in the History of Medicine and Jesus Research* 24). Its identification as a medical document was challenged by Joseph Naveh, "A Medical Document or a Writing Exercise? The So-Called 4Q Therapeia," *Israel Exploration Journal* 36 (1986): 52–5, and Charlesworth, after a close personal examination of the text, withdrew his assertion that it is a medical text (J. H. Charlesworth, "A Misunderstood Recently Published Dead Sea Scroll," *Explorations* 1, no. 2 [1994]: 2).

68. On Jewish physicians in the Roman Empire, see F. Kudlien, "Judische Ärzte im römischen Reich," *Medizinhistorisches Journal* 20 (1985): 36–57. According to the Talmud (Sanhedrin 17B), one of the ten requirements of the good life in a Jewish community was a surgeon (V. Nutton, "Continuity or Rediscovery? The City Physician in Classical Antiquity and Mediaeval Italy," in *The Town and State Physician in Europe from the Middle Ages to the Enlightenment*, ed. A. W. Russell, 23).

69. J. Jeremias, *Jerusalem in the Time of Jesus: An Investigation into Economic and Social Conditions during the New Testament Period* 17–18.

70. "Among early Christians," writes Fridolf Kudlien, "one will find some people whose hostility against physicians and medicine is almost outrageous" (F. Kudlien, "Cynicism and Medicine," *BHM* 48 [1974]: 317). Kudlien singles out four in particular as being categorically opposed to the use of medicine in any form: Marcion, Arnobius, Tatian, and Tertullian. Karl-Heinz Leven similarly counts them among those church fathers who "allowed only one kind of remedy, and that was prayer" (K.-H. Leven, *Medizinische bei Eusebios von Kaisareia* 62).

71. On the general acceptance of Greek medicine by the church fathers, see H.-J. Frings,

"Medizin und Arzt bei den griechischen Kirchernvätern bis Chrysostomos" 8–24; on Clement of Alexandria, see Dörnemann, *Krankheit und Heilung* 101–21.

72. See W. Jaeger, *Early Christianity and Greek Paideia,* esp. 26–35.

73. See the detailed investigation by Robert Grant of the attitudes of second-century church fathers toward natural philosophy (in which they had almost no interest with the exception of the dynamistic Monarchians, who were influenced by the writings of Galen) and of the fathers beginning in the third century, whom he finds in some respects "far less credulous than their [pagan] contemporaries" (*Miracle and Natural Law in Graeco-Roman and Early Christian Thought* 87–121, esp. 103 and 120).

74. "And when we find many of the Fathers like Jerome, Ambrose, and even Augustine speaking of pagan learning and literature with contempt and hostility it is necessary to remind ourselves that this contempt arises not from lack of education or a barbarous indifference towards knowledge as such, but from the vigour with which these men were pursuing a new ideal of knowledge, working in the teeth of opposition for a reorientation of the entire structure of human thought" (R. G. Collingwood, *The Idea of History* 51).

75. D. W. Amundsen, "Medicine and Faith in Early Christianity," *BHM* 56 (1982): 326–50, reprinted in Amundsen 129–32.

76. *First Apology,* ch. 44.

77. *Second Apology* 2.13.

78. *The Prescription against Heretics* 7.

79. Lindberg, "Science and the Early Church" 25–6.

80. See G. Rialdi, *La medicina nella dottrina di Tertulliano,* and S. D'Irsay, "Patristic Medicine," *Annals of Medical History* 9 (1927): 367–8; Dörnemann, *Krankheit und Heilung* 161–72.

81. Cf. *Apologeticum ad nationes* 1.15, *Apology* 9.

82. On Tertullian's education, see T. D. Barnes, *Tertullian: A Historical and Literary Study* 194–210; in medicine 205.

83. Ibid. 123; cf. 29, 198.

84. *De corona* 8.

85. Cf *Ad nationes* 2.4, *De corona* 8.

86. See D. G. Bostock, "Medical Theory and Theology in Origen," in *Origeniana Tertia: The Third International Colloquium for Origen Studies* 191–9; M. Dörnemann, "Medizinale Inhalte in der Theologie des Origenes," in *Ärztekunst und Gottvertrauen*: *Antike und mittelalterliche Schnittpunkte von Christentum und Medizin,* ed. C. Schulze and S. Ihm, 9–39; and Dörnemann, *Krankheit und Heilung* 121–60.

87. Homily 8 on Numbers 3.

88. *Adnotationes in librum III regum* 15:23.

89. *Contra Celsum* 3.12.

90. *Contra Celsum* 8.60.

91. Amundsen, "Medicine and Faith in Early Christianity," in Amundsen 140.

92. For examples, see Theodoret, *Historia Religiosa* 9.7 (Peter of Galatia) and Pseudo-Macarius, Homily 48.4 (both are quoted by Merideth 77–78).

93. D. W. Amundsen, "Body, Soul, and Physician," in Amundsen 26 n. 7.

94. See ch. 3, p. 52.

95. Glen Bowersock argues that there developed in the second century a taste for medicine that he attributes to the traditionally close association of medicine, philosophy, and oratory. Just as the emergence of Hippocratic medicine coincided with the sophistic movement in the fifth century B.C., so the "Second Sophistic" was accompanied by a new interest in medicine (*Greek Sophists in the Roman Empire* 66). P. A. Brunt convincingly rejects this widely accepted theory on the grounds that it confuses rhetoric with sophistic and that while there existed a close connection between medicine and philosophy, there was never a similar link between medicine and rhetoric ("The Bubble of the Second Sophistic," *Bulletin of the Institute of Classical Studies* [1994]: 37–46).

96. S. D'Irsay, "Christian Medicine and Science in the Third Century," *Journal of Religion* 10 (1930): 531–3.

97. Bostock, "Medical Theory and Theology in Origen" 191.

98. Ibid., 192–3; Origen, *Commentary on Matthew* 13, 4.

99. Bostock, "Medical Theory and Theology in Origen" 194.

100. Ibid., 197.

101. R. M. Grant, *Greek Apologists of the Second Century* 109–10; idem, *Miracle and Natural Law* 241–3; L. W. Bernard, "Athenagoras: De Resurrectione. The Background and Theology of a Second Century Treatise on the Resurrection," *Studia Theologica* 30 (1976): 11–16.

102. Grant, *Miracle and Natural Law* 235–45.

103. See S. R. Holman, *The Hungry Are Dying: Beggars and Bishops in Roman Cappadocia* 158–60. "Thus the symbolic image of contagion is here [in the discussion of the theory of airborne disease by Gregory of Nyssa and Gregory of Nazianzus] consistently rooted in a medical perception of the physical body" (ibid., 160).

104. Gregory of Nazianzus, *Or.* 43.23; see Holman, *The Hungry Are Dying* 29–30, and M. E. Keenan, "St. Gregory of Nazianzus and Early Byzantine Medicine," *BHM* 9 (1941): 12–14. Basil's brother, Gregory of Nyssa, though he had not studied medicine, had a deep interest in it (see M. E. Keenan, "St. Gregory of Nyssa and the Medical Profession," *BHM* 15 [1944]: 150–61).

105. J. P. Cavarnos, "Relation of the Body and Soul in the Thought of Gregory of Nyssa," in *Gregor von Nyssa und die Philosophie: Zweites Internationales Kolloquium über Gregor von Nyssa* 62.

106. Ibid., 71; cf. 72, 73, 74; cf. Merideth 32–3.

107. Bernard, "Athenagoras" 2–3.

108. L. Edelstein, "The Relation of Ancient Philosophy to Medicine," in Edelstein 360–66; J. Pigeaud, *La Maladie de l'ame: Etude sur la relation de l'âme et du corps dans la tradition médico-philosophique antique.*

109. For a detailed discussion see M. C. Nussbaum, *The Therapy of Desire: Theory and Practice in Hellenistic Ethics* 13–47 and passim, and, for a more specific discussion, G. Ferngren and D. W. Amundsen, "Virtue and Health/Medicine in Pre-Christian Antiquity," in *Virtue and Medicine: Explorations in the Character of Medicine*, ed. E. E. Shelp, 9–11.

110. *Republic* 443D–E. On Plato, see Lloyd, *In the Grip of Disease* 142–57, and J. T. McNeill, *A History of the Cure of Souls* 17–41.

111. Plato, *Laws* 862C. In contrast, while physical healing is employed figuratively to describe Yahweh's salvation in the Hebrew scriptures, sin is never treated as a spiritual sickness (*TDNT* 3:203, s.v. ἰάομαι).

112. B. Snell, *The Discovery of the Mind* 162.

113. Ἰατρεῖόν ἐστιν, ἄνδρες, τὸ τοῦ φιλοσόφου σχολεῖον· οὐ δεῖ ἡσθέντας ἐξελθεῖν, ἀλλ' ἀλγήσαντας. ἔρχεσθε γὰρ οὐχ ὑγιεῖς, ἀλλ' ὁ μὲν ὦμον ἐκβεβληκώς, ὁδ' ἀπόστημα ἔχων, ὁ δὲ σύριγγα ὁ δὲ κεφαλαλγῶν. εἶτ' ἐγὼ καθίσας ὑμῖν λέγω νοημάτια καὶ ἐπιφωνημάτια, ἵν' ὑμεῖς ἐπαινέσαντές με ἐξέλθητε, ὁ μὲν τὸν ὦμον ἐκφέρων οἷον εἰσήνεγκεν, ὁ δὲ τὴν κεφαλὴν ὡσαύτως ἔχουσαν, ὁ δὲ τὴν σύριγγα, ὁ δέ τὸ ἀπόστημα; εἶτα τούτου ἕνεκα ἀποδημήσωσιν ἄνθρωποι νεώτεροι καὶ τοὺς γονεῖς τοὺς αὑτῶν ἀπολίπωσιν καὶ τοὺς φίλους καὶ τοὺς συγγενεῖς καὶ τὸ κτησίδιον, ἵνα σοι "οὐᾶ" φῶσιν ἐπιφωνημάτια λέγοντι; τοῦτο Σωκράτης ἐποίει, τοῦτο Ζήνων, τοῦτο Κλεάνθης; (Epictetus, *Discourses* 3.23.30–32, Matheson translation, in *The Stoic and Epicurean Philosophers,* ed. W. J. Oates).

114. H. Schipperges, "Zur Tradition des 'Christus Medicus' im frühen Christentum und in der alteren Heilkunde," *Arzt und Christ* 11 (1965): 12–20; G. Fichtner, "Christus als Artzt: Ursprünge und Wirkungen eines Motivs," *Frühmittelalterliche Studien* 16 (1982): 1–18; Merideth 157–62; M. Dörnemann, *Krankheit und Heilung* 88–100; and C. Schulze, *Medizin und Christentum in Spatäntike und frühem Mittelalter: Christliche Ärzte und ihr Wirk* 156–63.

115. P. Cordes, *Iatros: Das Bild des Arztes in der Griechischen Literatur von Homer bis Aristoteles.*

116. G. Dumeige, SJ, "Le Christ médecin dans la littérature chrétienne des premiers siècles," *Rivista di archeologia cristiana* 47 (1972): 115–41.

117. *TDNT* 3:214–15, s.v. ἰάομαι; A. S. Pease, "Medical Allusions in the Works of St. Jerome," *Harvard Studies in Classical Philology* 25 (1914): 74–5, 78; P. C. J. Eijkenboom, *Het Christus-Medicusmotief in de preken van Sint Augustinus* (Christus medicus in the sermons of St. Augustine) 222–4. Morton Smith observes that ἴασις and ἰάομαι are normally used of the healing of *sins* in the Apostolic Fathers ("De tuenda sanitate praecepta (Moralia, 122B–137E)," in *Plutarch's Ethical Writings and Early Christian Literature,* ed. H. D. Betz, 37 and n. 19 for citations).

118. οἷά τις ἰατρῶν ἄριστος τῆς τῶν καμνόντων ἕνεκεν σωτηρίας 'ὁρῇ μὲν δεινά, θιγγάνει δ' ἀηδέων ἐπ' ἀλλοτρίῃσί τε ξυμφοῇσιν ἰδίας καρποῦται λύπας' (Περὶ φυσῶν [*On Breaths*] I. 2 ap. *Eccles. Hist.* 10.4.11). For a critical text, translation, and testimonia, see J. Jouanna, *Hippocrate, Les Vents, De l'Art.*

119. Pace E. J. Edelstein and L. Edelstein, *Asclepius: A Collection and Interpretation of the Testimonies* 2:135 n. 10. Cf. the language of Origen, Homily 1 on Psalm 37. This familiar Hippocratic aphorism was, however, popular among Christians in describing the compassion of physicians (Merideth 143–4), finding echoes in Origen, Homily on Jeremiah 14 (trans. O. Temkin, *Hippocrates in a World of Pagans and Christians* 143), and Gregory of Nazianzus, *On Basil* 35. On Eusebius's metaphorical use of medical terms, see Dörnemann, *Krankheit und Heilung* 180–9.

120. An extensive literature exists on the cult of Asclepius. In particular, see Edelstein and Edelstein, *Asclepius;* A. Krug, *Heilkunst und Heilkult: Medizin in der Antike* 120–87 (for an overview of major sanctuaries and healing sites in the Western Empire, see 172–81); F. Steger,

Asklepiosmedizin: Medizinischer Alltag in der römischen Kaiserzeit; and J. W. Riethmüller, *Asklepios: Heiligtümer und Kulte,* 2 vols. On Asclepius as a competitor to Jesus see Edelstein and Edelstein, *Asclepius* 2:132–8.

121. Ἔστιν οὖω ὁ παιδαγωγὸς ἡμῶν λόγος διὰ παραινέσεων θεραπευτικὸς τῶν παρὰ φύσιν τῆς ψυχῆς παθων. Κυρίως μὲν γὰρ ἡ τῶν τοῦ σώματος νοσημάτων βοήθεια ἰατρικὴ καλεῖται τέχνη ἀνθρωπίνῃ σοφία διδακτή. Λόγος δὲ ὁ πατρικὸς μόνος ἐστὶν ἀνθρωπίνων ἰατρὸς ἀρρωστημάτων παιώνιος καὶ ἐπῳδὸς ἅγιος νοσούσης ψυχῆς (*Paedagogos* 1.2.6). The context makes it clear that the reference to "human infirmities" (ἀνθρωπίνων ἀρρωστημάτων) is intended to be understood in a metaphorical sense only.

122. V. Nutton, "Murders and Miracles," in *From Democedes to Harvey* VIII 50 n. 90. Nutton writes that "his hostility was also sharpened by the continuing paganism of many intellectuals."

123. Amundsen, "Medicine and Faith," in Amundsen 150.

124. See J. Quasten, *Patrology* 2:383–92 (for detailed bibliographies, 2:386–7, 392); P. de Labriolle, *History and Literature of Christianity from Tertullian to Boethius,* trans. H. Wilson, 188–99; C. T. Cruttwell, *A Literary History of Early Christianity* 2:630–42; A. Di Berardino, ed., *Encyclopedia of the Early Church,* trans. A. Walford, 1:82, s.v. "Arnobius of Sicca" (with extensive bibliography); M. B. Simmons, *Arnobius of Sicca: Religious Conflict and Competition in the Age of Diocletian.*

125. Four passing references in various letters, and one mention each in his continuation of Eusebius's *Chronicle,* in the life of Lactantius in *De viris illustribus,* and in his own life in *De viris illustribus* (O. P. Nicholson, "The Date of Arnobius' *Adversus gentes,*" *Studia Patristica* 15 [1984]: 101).

126. Arnobius of Sicca, *The Case against the Pagans,* 2 vols., newly translated and annotated by G. E. McCracken. All quotations are from this translation.

127. The evidence for assigning dates to the events of his life is almost entirely circumstantial. The date of his conversion is given by Jerome as 327, but the internal evidence of the *Adversus nationes* seems to suggest an earlier date, during the first decade of the fourth century, either before or during the persecution of Diocletian. We cannot be more precise than this.

128. With the aid of Lucille Berkowitz's *Index Arnobianus* it is a relatively easy task to find every mention of disease, medicine, healing, and physicians in the *Adversus nationes.*

129. They are, says Arnobius, ascribed to Aesculapius, who discovered medicine (1.41.4).

130. Arnobius is not expressing a suspicion of physicians for the medicine they practiced. When he speaks of their "despising those in which but now they trusted," Arnobius has in mind their former (i.e., pagan) beliefs, not the arts that they practiced.

131. It was unusual to ascribe disease to the Fates. Arnobius was, I think, merely personifying one's destiny as divinely sent or appointed. He probably referred to illnesses that fall to one's lot in life. Tatian also believed in the reality of fate but held that Christians are above it (D. W. Amundsen, "Tatian's 'Rejection' of Medicine," in Amundsen 161).

132. "Murders and Miracles" VIII 46. See H. F. J. Horstmanshoff, " 'Did the God Learn Medicine?' Asclepius and Temple Medicine in Aelius Aristides' *Sacred Tales,*" in *Magic and Rationality in Ancient Near Eastern and Graeco-Roman Medicine,* ed. H. F. J. Horstmanshoff and M. Stol, 325–41; cf. Edelstein and Edelstein, *Asclepius* 2:112 n. 4, and Lloyd, *Magic, Reason and Experience* 41 (on Aelius Aristides).

133. The charge was made by other critics of Asclepius. Diodorus Siculus (first century B.C.) writes that Asclepius not only healed people who were thought to be beyond help but also brought the dead to life. He did so, however, not in a wondrous fashion but by using his medical knowledge (4.71.1; see Remus, *Pagan-Christian Conflict* 37).

134. "As a man of science, his chief defect is an inability to grasp the difference between problems that are really beyond the reach of the human mind, and such as, like those of physical science, are discoverable by the use of a true method" (Cruttwell, *A Literary History of Early Christianity* 2:641; cf. 638).

135. See Edelstein and Edelstein, *Asclepius* 2:255.

136. Merideth 73–4. Scholars also, Merideth observes, selectively quote those authors who support their opinions regarding Christian attitudes to medicine.

137. Ibid., 75. Merideth believes that Tatian denigrated medicine as part of a larger rejection of classical culture.

138. Ibid., 63, quoting David Westerlund.

139. Ibid., 63.

140. Ibid., 71.

141. P. Brown, *The Cult of the Saints: Its Rise and Function in Latin Christianity* 114–15.

142. P. Horden, "Saints and Doctors in the Early Byzantine Empire: The Case of Theodore of Sykeon," in *The Church and Healing,* ed. W. J. Sheils, 12–13.

143. G. E. R. Lloyd, *Demystifying Mentalities* 30–31; cf. Lloyd, *Magic, Reason and Experience* 38, where he adds gymnastic trainers to the list.

144. Nutton, "Murders and Miracles" VIII 40. To these lists Rebecca Flemming adds magicians (*magi*), astrologers (*mathematici*), dream interpreters (*oneirokritai*), and old women (*aniles*) (*Medicine and the Making of Roman Women* 33).

145. Flemming, *Medicine and the Making of Roman Women* 35–44.

146. See, e.g., ibid., 77. While there certainly existed a "range of curative possibilities" in the medical pluralism of Roman imperial society (ibid., 73), the evidence contradicts her assertion that no type of health care could claim ascendancy.

147. Nutton, "Roman Medicine" 56.

148. V. Nutton, "Beyond the Hippocratic Oath," in *Doctors and Ethics: The Earlier Historical Setting of Professional Ethics,* ed. A. Wear, J. Geyer-Kordesch, and Roger French, 15–16.

149. Nutton, "Murders and Miracles" VIII 33.

150. Nutton, "Continuity or Rediscovery?" 18–19.

151. Ibid., 9 and 35 n. 6.

152. Nutton, "Roman Medicine" 69.

153. See Galen's description in *De part. art. Medic.* 2, which is quoted in translation by Nutton, "Roman Medicine" 64–5.

154. Ibid., 72–3 and n. 104, where Nutton cites inscriptional evidence.

155. Ibid., 76 and n. 116 for the citation.

156. Ibid., 67.

157. Ibid., 69.

158. M. I. Finley, *Ancient Slavery and Modern Ideology* 105–7.

159. Rousselle, "From Sanctuary to Miracle-Worker" 97. Rousselle's reconstruction is speculative.

160. Nutton, "Roman Medicine" 71.

161. Most ancient literature was written by members of the upper classes (see M. Smith, "Prolegomena to a Discussion of Aretalogies, Divine Men, the Gospels and Jesus," *Journal of Biblical Literature* 90 [1971]: 179).

162. Flemming, *Medicine and the Making of Roman Women* 77.

163. Flemming admits that folk medicine "falls short of the threshold of historical visibility" (ibid., 82).

164. Morton Smith terms the New Testament "a lower-middle-class product" (Smith, "Prolegomena" 179).

165. Merideth 84–5; Horden, "Saints and Doctors in the Early Byzantine Empire" 10.

166. R. Browning, "The 'Low Level' Saint's Life in the Early Byzantine World," in *The Byzantine Saint,* ed. S. Hackel, 122–3.

167. V. Nutton, review of S. Kottek, M. Horstmanshoff et al., *From Athens to Jerusalem: Medicine in Hellenized Jewish Lore and in Early Christian Literature, BHM* 75 (2001): 787–8.

168. It might seem as if early Christians, who before 313 were undergoing persecution by Roman authorities, would be marginalized in a pagan society and hence likely to adopt sectarian attitudes in a variety of cultural matters, including medicine. But Stark (59) points out that Christianity presented an outward face to Graeco-Roman culture (and another to hellenized Jews) that permitted it both to reach out to that culture and to accommodate itself in many ways to it.

169. Ambrose, *On Cain* 1.40; Jerome, *On Isaiah* 8; and Augustine, Tractate 30 on John 3; *On Christian Doctrine* 4.16.33; Sermon 80.3 and Sermon 84, in Augustine, *Sermons,* in *The Works of St. Augustine: A Translation for the 21st Century,* ed. J. E. Rotelle, OSA, and trans. E. Hill, OP.

170. In late antiquity pagan healing practices, and incubation in particular, were adopted by some Christians. See M. Hamilton, *Incubation: Or, the Cure of Disease in Pagan Temples and Christian Churches* 109–71, and I. Csepregi, "The Compositional History of Greek Christian Incubation Miracle Collections: Saint Thecla, Saint Cosmas and Damian, Saint Cyrus and John, Saint Artemios."

Chapter 3 • Early Christian Views of the Etiology of Disease

1. For a brief but comprehensive account of demonology in the classical world, see Luck, *Arcana Mundi* 163–75; a detailed survey of current scholarship is found in F. E. Brenk, SJ, "In the Light of the Moon: Demonology in the Early Imperial Period," in *ANRW* II. 16, 3 (1986): 2068–145.

2. See, e.g., S. J. Case, "The Art of Healing in Early Christian times," *Journal of Religion* 3 (1923): 238–55, esp. 253–5; Nutton, "Murders and Miracles" VIII 48; E. H. Ackerknecht, *A Short History of Medicine,* rev. ed., 81; L. Edelstein, "Greek Medicine in Its Relation to Religion and Magic" in Edelstein 222 and n. 53; M. Smith, "De tuenda sanitate praecepta" 35–36. Morton Smith thinks that early Christian authors reveal "the almost total absence of rational medical advice," though they show knowledge of common medical practices of the period and use medical commonplaces and metaphors (36–7). He attributes this absence to the belief that medicine was not worthy of concern for Christians because of "a strong

tradition that the body is contemptible and to be neglected" and the belief that miraculous healing made it unnecessary (40, 41). Dale Martin holds that early Christians generally believed that disease was caused by demons, but he believes that another, noninvasive view of disease can be detected in the apostles John and Paul (*The Corinthian Body* 164–5). One even finds demons defined by Arndt and Gingrich, as those "who are said to enter into persons and cause illness, especially of the mental variety. . . . Hence the healing of a sick person is described as the driving out of demons" (W. F. Arndt and F. W. Gingrich, *A Greek-English Lexicon of the New Testament and Other Early Christian Literature* 168, s.v. δαιμόνιον).

3. Seybold and Mueller, *Sickness and Healing* 100.

4. O. Böcher, *Dämonenfurcht und Dämonenabwehr. Ein Beitrag zur Vorgeschichte der christlichen Taufe*; *Das neue Testament und die dämonischen Mächte*; and *Christus Exorcista. Damonismus und Taufe im Neuen Testament.* Böcher's views are followed by E. A. Leeper, "Exorcism in Early Christianity," who assumes rather than demonstrates that demons were widely thought to cause disease in the first century (91–3).

5. E. Yamauchi, "Magic or Miracle? Diseases, Demons and Exorcism," in *The Miracles of Jesus*, ed. D. Wenham and C. Blomberg, 92.

6. See, e.g., Foerster's remark: "it should be noted that in the NT not all sicknesses are attributed to demons even in older strata of the Synoptic tradition" (*TDNT* 2:18, s.v. δαίμων κτλ).

7. L. D. Hankoff, "Religious Healing in First-Century Christianity," *Journal of Psychohistory* 19, no. 4 (1992): 387–407.

8. Ibid., 393–4.

9. Yamauchi, "Magic or Miracle?" 99.

10. J. B. Russell, *The Devil: Perceptions of Evil from Antiquity to Primitive Christianity* 217–20.

11. No instance of exorcism is described in the Old Testament. In spite of statements to the contrary, 1 Samuel 16:14–23 does not describe exorcism. On the date and background of Tobit, see D. A. deSilva, *Introducing the Apocrypha: Message, Context, and Significance* 63–84, and L. P. Hogan, *Healing in the Second Tempel* [*sic*] *Period* 27–37. On Tobit's demonology, see B. Kollmann, *Jesus und die Christen als Wundertäter*: *Studien zu Magie, Medizin und Schamanismus in Antike und Christentum* 120–4.

12. Kollmann, *Jesus und die Christen als Wundertäter* 137–54.

13. *Antiquities* 8.2.5 (45); Kollmann, *Jesus und die Christen als Wundertäter* 147–51.

14. Josephus, *Antiquities* 8.2.5 (45); S. S. Kottek, *Medicine and Hygiene in the Works of Flavius Josephus* 16–18 and passim.

15. Josephus, *War* 2.8. 6 (136); Kollmann, *Jesus und die Christen als Wundertäter* 127–31; Kottek, *Medicine and Hygiene* 161–70; Hogan, *Healing in the Second Tempel Period* 136–67. Hogan is cautious about identifying the Qumran community with the Essenes (136–7).

16. On exorcism in the Qumran literature, see Kollmann, *Jesus und die Christen als Wundertäter* 131–7.

17. Ibid., 156–60.

18. On differences between Jesus, exorcists, and magicians, see D. W. Amundsen and G. B. Ferngren, "The Healing Miracles of the Gospels: Problems and Methods," in *ANRW* II. 26, 3 (forthcoming). For the view that Jesus was a wonder-worker or magician on the Hellenistic model, see M. Smith, *Jesus the Magician;* J. M. Hull, *Hellenistic Magic and the*

Synoptic Tradition; and G. Vermes, *Jesus the Jew: A Historian's Reading of the Gospels;* contra Aune ("Magic in Early Christianity" 1539), who does not believe that one should categorize Jesus as a magician. Elizabeth Ann Leeper rejects the views of Smith and Hull but thinks that Jesus saw himself as an exorcist and was viewed as one ("Exorcism in Early Christianity" 75–82). On the apologists' distinguishing between the miracles of Jesus and charges of magic, see Achtemeier, "Jesus and the Disciples as Miracle Workers" 159–60.

19. G. E. Ladd, *The Presence of the Future: The Eschatology of Biblical Realism* 149–50. See Lk. 4:33–34 and 41; cf. Mk. 1:32–34 and 39.

20. ἐξέβαλεν τὰ πνεύματα λόγῳ καὶ πάντας τοὺς κακῶς ἔχοντας ἐθεράπευσεν (Mt. 8:16; cf. Mk. 6:12–13 and Acts 19:12). See S. Eitrem, *Some Notes on the Demonology in the New Testament* 28. Géza Vermès contrasts healing and exorcism in the Gospels, in which they are separated, with the Genesis Apocryphon (1QGenAp) found at Qumran, in which they form a single process (*Jesus the Jew* 66).

21. See Amundsen and Ferngren, "Perception of Disease" 2949–55, for a fuller discussion.

22. The concept of healing here is that of Jesus's making a man whole.

23. While it is possible that the possession and impairment are unrelated and only fortuitously found in the same person, it is a more economical hypothesis that the impairments are attributed to the demons. In Luke's narrative the demon is described as a "mute demon," that is, a demon who causes muteness.

24. The case of a woman who endured a bent spine and suffered from "a spirit that had crippled her" does not, in my opinion, belong in this category (see below, n. 126).

25. Cf. Aune, "Magic in Early Christianity" 1529.

26. Case, "The Art of Healing" 253.

27. Nutton, "Murders and Miracles" VIII 46.

28. πολλὰ παθοῦσα ὑπὸ πολλῶν ἰατρῶν καὶ δαπανήσασα τὰ παρ' αὐτῆς πάντα (Mk. 5:24–34, quotation at v. 26). The parallel passage in Luke 8:43b is a late interpolation.

29. Horstmanshoff, " 'Did the God Learn Medicine?' " 328–9 n. 10, with many citations.

30. The expression "signs and wonders" (σημεῖα καὶ τέρατα) is used of Paul and Barnabus at Iconium (Acts 14.3) and "signs" (τὰ σημεῖα) of Philip in Samaria (Acts 8:6).

31. See Acts 5:16 and 8:7.

32. τά τε πνεύματα τὰ πονηρὰ ἐκπορεύεσθαι (Acts 19:11–12; cf 5:15–16 and 8:6–7). This formula may find an echo as well in Peter's words in Acts 10:38. Henry Kelly calls these words "perhaps the clearest statement of the synoptic tendency to attribute all illness to the devil" (*The Devil, Demonology and Witchcraft* 70). But see below pp. 59–61.

33. The situation in the Gospels in quite different. In one instance Jesus refuses to accede to the request of his disciples that he condemn one who casts out demons in his name (Jn. 9:38–40).

34. See J. A. Hardon, "The Miracle Narratives in the Acts of the Apostles," *CBQ* 16 (1954): 303–5.

35. See *TDNT* 2:16, s.v. δαίμων κτλ.

36. A. Harnack, *The Mission and Expansion of Christianity in the First Three Centuries*, trans. and ed. J. Moffatt, 1:162 n. 1.

37. Cf. Aune, "Magic in Early Christianity" 1548.

38. See *TDNT* 2:16–17, s.v. δαίμων κτλ.

39. Aune, "Magic in Early Christianity" 1548. Paul's "thorn in the flesh" (2 Cor. 12:7–10) remained unhealed, and his infirmity required care (see ch. 4, n. 23); Paul's friend Epaphroditus recovered, but without indication in the text of miraculous healing (Phil. 2:25–27); Paul left his companion Trophimus ill in Miletus (2 Tim. 4:20).

40. See Mk. 1:32–34 (= Mt. 8:16 = Lk 4:40–41), 1:39 (= Mt. 4:24), 3:10–11 (= Lk. 6:17–18), Lk. 13:32, and Aune, "Magic in Early Christianity" 1529 and n. 96, to whom I am indebted for these citations.

41. Brenk, "In the Light of the Moon" 2068–91 on Greek conceptions of demons from Homer through the philosophers, 2094–8 on the Neopythagoreans, 2098–107 on Philo, and 2117–42 on Plutarch, Lucian, and Apuleius; E. Ferguson, *Backgrounds of Early Christianity* 184–6. For a detailed account of the development of pagan and Christian ideas of demonism in late antiquity, see V. Flint, "The Demonisation of Magic and Sorcery in Late Antiquity: Christian Redefinitions of Pagan Religions," in *Witchcraft and Magic in Europe: Ancient Greece and Rome*, by V. Flint, R. Gordon, G. Luck, and D. Ogden, 277–348. Flint overestimates the influence of demonic belief on pre-Constantinian Christianity.

42. Edelstein, "Greek Medicine in Its Relation to Religion and Magic" 219–26.

43. See ch. 2, p. 24.

44. Edelstein, "Greek Medicine in Its Relation to Religion and Magic" 220.

45. *Enneades* II.9.14. The passage is quoted by Edelstein in both Greek and English translation ("Greek Medicine in Its Relation to Religion and Magic" 221).

46. Edelstein, "Greek Medicine in Its Relation to Religion and Magic" 223.

47. Homily 8 on Colossians 62.357–8 as quoted by Merideth 97. Merideth places prayers, spells, and incantations in the category of "ritualized speech" and argues (wrongly, I believe) that they are not very different. Chrysostom's point, however, is that incantations have no part in the art of medicine or in a physician.

48. E. R. Dodds, *The Greeks and the Irrational* 252–3. Dodds maintains that the Greeks had retreated from an "open society," which demonstrated a "fear of freedom" that "marked a general change in the intellectual climate of the Mediterranean world" (248–9). In fact, the second century is not likely to have been any more superstitious than other centuries (Gager, *Curse Tablets and Binding Spells* 121–2). On *deisdaimonia,* see Ferguson, *Backgrounds of Early Christianity* 186.

49. *Soranus' Gynecology* I.II.4, translated with an introduction by O. Temkin, 4; cf. II.XII.19, p. 93. On Soranus's treatment of superstition, see ibid., xxxi–xxxii.

50. Martin, *The Corinthian Body* 152–3.

51. Leeper, "Exorcism in Early Christianity" 177.

52. Martin, *The Corinthian Body* 161–2. On the difficulty of assuming a correlation between class and interest in intellectual speculation, see ch. 4, p. 75.

53. Aune, "Magic in Early Christianity" 1531.

54. E. R. Dodds, *Pagan and Christian in an Age of Anxiety: Some Aspects of Religious Experience from Marcus Aurelius to Constantine* 117.

55. Edelstein, "Greek Medicine in Its Relation to Magic" 223–4. Edelstein believes that Christians and Jews spread ideas of demonic etiology that they took over initially from the Persians and the Chaldeans (222 and n. 53).

56. The Fourth Gospel and the Apocalypse are regarded by the majority of scholars as having been written in the last decade of the first century (see D. Guthrie, *New Testament Introduction* 282–3 and 949).

57. See W. M. Alexander, *Demonic Possession in the New Testament: Its Historical, Medical, and Theological Aspects* 216–21. Ignatius's Epistle to the Antiochians speaks of an order of exorcists, but the epistle is regarded as spurious (219–20 n. 2).

58. *Second Apology* 6. For a discussion of the demonology of the second-century apologists, see E. Ferguson, *Demonology of the Early Christian World* 105–42.

59. *Second Apology* 6. Most of his references to healing are ambiguous (e.g., *Dialogue with Trypho* 30, 39, 76, 85). They have been cited as evidence of physical healing by M. T. Kelsey, *Healing and Christianity in Ancient Thought and Modern Times* 149 n. 23, but they probably refer to the expulsion of demons.

60. *Ad Scapulam* 2 (translation, *ANF* 3:106); cf. *Apology* 23 and 37.

61. Leeper, "Exorcism in Early Christianity" 153.

62. See Tertullian, *Apology* 22; Origen, *Contra Celsum* 1.31 and 8.31.

63. Origen believed, as did most Christians, that Asclepius was a demon who had the power to heal *(Contra Celsum* 3.35).

64. *Octavius* 27 (translation, *ANF* 4:190). It is perhaps out of these beliefs that accedie (*akêdia*) developed among monastic writers in late antiquity. Accedie (sloth) was believed to be a false illness that was caused by demons, and it was regarded as a common temptation among monastics for which they bore personal responsibility. It produced both physiological and psychological symptoms (see Crislip 78–81 and 180 n. 61 for a survey of the literature).

65. On Tatian's view of medicine, see D. W. Amundsen, "Tatian's 'Rejection' of Medicine in the Second Century," in *Ancient Medicine in Its Socio-Cultural Context,* ed. P. J. van der Eijk et al., 2:377–92 (reprinted in Amundsen 158–74), and Temkin, *Hippocrates in a World of Pagans and Christians* 119–23.

66. Kudlien, "Cynicism and Medicine" 318. On Kudlien's misunderstanding of Tatian, see Amundsen 10.

67. E.g., Tatian writes that demons induce disturbances in men, but if they are confronted by God's word, they leave them, and the sick persons are healed (*Oration to the Greeks* 16). Vivian Nutton overlooks the ambiguity in Tatian when he attributes to him the view that demons are the true cause of disease ("Murders and Miracles" VIII 50).

68. Tatian, *Oration to the Greeks* 18 (translation, *ANF* 2:73).

69. "Very kind, too, no doubt, they are in regard to the healing of diseases. For, first of all, they make you ill; then, to get a miracle out of it, they command the application of remedies either altogether new, or contrary to those in use, and straightaway withdrawing hurtful influence, they are supposed to have wrought a cure" (*Apology* 22; translation, *ANF* 3:37).

70. *Against Heresies* 2.32.4; cf. 2.31.2. On these passages see ch. 4.

71. See Warfield 11–16.

72. *Contra Celsum* 1.46. On this passage, see Warfield 239–40 n. 22.

73. Leeper, "Exorcism in Early Christianity" 158–9.

74. But some were: see Lane Fox, *Pagans and Christians* 329.

75. Leeper, "Exorcism in Early Christianity" 158.

76. See ch. 2, pp. 25–29, on the apologists' attitudes to medicine. Of course, a high regard

for medicine does not in itself ensure acceptance of the principle of natural causation, but familiarity with natural processes is likely to incline the mind in that direction.

77. Leeper, "Exorcism in Early Christianity" 177–9, to whose discussion and survey of the literature I am indebted.

78. See, e.g., M. P. Nilsson, *Greek Piety* 170–1; Dodds, *The Greeks and the Irrational* 133–8.

79. P. Brown, *The World of Late Antiquity* 49–57; cf. Harnack, *The Expansion of Christianity* 1:152–80. For a detailed survey of literature on demonology in the ancient world, particularly as it relates to healing, see Yamauchi, "Magic or Miracles?" 89–183.

80. See Warfield 12; Alexander, *Demonic Possession* 221–33.

81. See ch. 4, p. 76, and n. 99.

82. A. D. Nock, *Conversion: The Old and the New in Religion from Alexander the Great to Augustine of Hippo* 104–5. Nock cites Celsus as evidence that Christian exorcism for healing "impressed the popular imagination" and thinks Christianity unique in making exorcism an officially sanctioned activity (104).

83. J. G. Gager, *Kingdom and Community* 140–1. The evidence suggests that exorcisms—pagan, Jewish, Christian—were rarer than some modern scholars suggest.

84. Brown, *The World of Late Antiquity* 55; see Merideth 17.

85. Leeper, "Exorcism in Early Christianity" 177–9.

86. See above, p. 54 and n. 41 (on evidence from Plotinus).

87. Rowan Greer maintains that this theme summarizes the message of Christianity during the patristic period (*The Fear of Freedom: A Study of Miracles in the Roman Imperial Church* 81).

88. H. E. W. Turner, *The Patristic Doctrine of Redemption: A Study of the Development of Doctrine during the First Five Centuries* 47–69; C. Brown, ed., *The New International Dictionary of New Testament Theology* 1:644–52, s.v. "Fight et al.," by Günther. For citations to the patristic sources, see Turner's comprehensive treatment.

89. See, e.g., Irenaeus: "ut occideret peccatum, evacuaret autem mortem, et vivificaret hominem" (*Adv. Haer. Omn.* 3.19.6; quoted by Turner, *The Patristic Doctrine of Redemption* 51 n. 5).

90. M. Green, *Evangelism in the Early Church* 190.

91. Alexander, *Demonic Possession* 233.

92. Hippolytus, *Apostolic Tradition* 21 (G. J. Cuming, *Hippolytus: A Text for Students with Introduction, Translation, Commentary and Notes* 18–19).

93. *DCA* 2:650–3, s.v. "Exorcism." This multitude of liturgical exorcisms might, however, be the source of the accounts of numerous exorcisms referred to by contemporary writers.

94. G. W. H. Lampe, "Miracles and Early Christian Apologetic," in Moule, *Miracles* 215.

95. See, e.g., Cyril of Jerusalem, *Myst. Cat.* 1.4, and John Chrysostom, *Cat.* 2.18; J. B. Russell, *Satan: The Early Christian Tradition* 100–3.

96. According to Eusebius there were 52 exorcists in the church in Rome in the mid-third century (*Eccles. Hist.* 6.43).

97. See, e.g., Tertullian, *Ad Scapulam* 4; Lampe, "Miracles and Early Christian Apologetic" 215–18 on exorcism; impression made on pagans at 217.

98. On the ubiquity of magic in late antiquity, which was lacking in classical and Hellenistic times, see Nilsson, *Greek Piety* 162–3.

99. *City of God* 22.8. On Augustine's changing views regarding miracles, see ch. 4, p. 78.

100. The involvement of demons in book 22 is limited to two cases. In one case a physician with gout was plagued by "black woolly-haired boys whom he took to be demons" (*a pueris nigris cirratis quos intellegebat daemones*). They appeared in dreams to inflict pain by stepping on his feet. In a second (that of Hesperius), Hesperius's family, cattle, and servants were plagued by demons, from which they were delivered by prayer. In neither case is a disease or impairment attributed to demons.

101. In monastic health care, diagnostic procedures required distinguishing between afflictions (physical, mental, spiritual) to determine, among other things, whether illnesses were the result of demonic affliction or natural causes. *Diakrisis* ("discernment") was the method of the determination, but it was a component that went beyond what we should today call diagnosis. It was a skill that the elders of the monastic community alone possessed, and it required more than medical knowledge; it demanded as well divine illumination, and it was more than a matter of the mere determination of disease etiology. When a monk complained of sickness, the elders assessed whether his illness was genuine or pretended, and they determined the treatment (e.g., dietary therapy) in each case (see Crislip 18–21). "The taxonomy of demons, and the identification of their influence upon monastics, is perhaps the most common use of discernment in monastic literature" (ibid., 19). But demonic affliction was not limited to illness: demons afflicted one's thoughts and passions as well. Moreover, the prescribed therapy for a demonic and a nondemonic illness (e.g., blessed oil, the sign of the cross) was not always different (ibid., 22–23, 25). Prayer, exorcism, invocation of Christ's name, and the mere presence of a monastic could cure demonically attributed diseases (ibid., 25).

102. See O. Temkin, *The Falling Sickness: A History of Epilepsy from the Greeks to the Beginnings of Modern Neurology*, 2d ed., rev., 51–64; H. W. Miller, "The Concept of the Divine in *De Morbo Sacro*," *TAPA* 84 (1953): 1–15.

103. Temkin follows F. J. Dölger, who suggests that the explanation of epilepsy as demonic possession during the Middle Ages was largely due to Origen's influence (F. J. Dölger, "Der Einfluss des Origenes auf die Beurteilung der Epilepsie und Mondsucht im christlichen Altertum," *Antike und Christentum* 4 [1934]: 95–109). Origen argues for a demonological explanation of the moonstruck boy in Mt. 17:15 against those physicians who would diagnose it as a physical condition (*Commentary on Matthew* 13, 4, col. 1104). Temkin cites no evidence in support of his view that "Origen's explanation was followed by many Greek and Latin Fathers of the Church" *(The Falling Sickness* 92). In fact, a number of them accepted physiological explanations of epilepsy: see, e.g., Clement of Alexandria, *Instructor* 10; *Stromata* 7; Tertullian, *Apology* 9.

104. S. B. Thielman and F. S. Thielman, "Constructing Religious Melancholy: Despair, Melancholia, and Spirituality in the Writings of the Church Fathers of Late Antiquity," paper presented at the sixty-ninth meeting of the American Association for the History of Medicine, Williamsburg, Va., April 4, 1997. For a survey of Greek medical theories regarding melancholy, see H. Flashar, *Melancholie und Melancholiker in den medizinischen Theorien der Antike.*

105. Jerome, Epistle 125.16; cf. *Augustine, De cura pro mortuis gerenda* 12.14.

106. P. Horden, "Responses to Possession and Insanity in the Earlier Byzantine World," *Social History of Medicine* 6 (1993): 177–94.

107. Ibid., 186–7.

108. Ibid., 191.

109. Edelstein, "Greek Medicine in Its Relation to Religion and Magic" 219–20 and n. 48.

110. Horden, "Responses" 181, 185.

111. Ibid., 179.

112. Thus Theodoret of Cyrrhus records the case of a delirious woman: Some "called it [her condition] the action of a demon, while the doctors named it a disease of the brain" (*Historia Religiosa* 13.13 [quoted by Merideth 59]).

113. Against the view that a naturalistic understanding of mental illness was replaced in late antiquity by a widespread belief in supernatural causes, which dominated the Middle Ages, see J. Kroll, "A Reappraisal of Psychiatry in the Middle Ages," *Archives of General Psychiatry* 29 (1973): 276–83, and J. Kroll and B. Bachrach, "Sin and Mental Illness in the Middle Ages," *Psychological Medicine* 14 (1984): 507–14. Kroll and Bachrach argue that a naturalistic etiology of disease in general was much more common in the Middle Ages than is usually assumed: see idem, "Sin and the Etiology of Disease in pre-Crusade Europe," *JHM* 41 (1986): 395–414.

114. Remus, *Pagan-Christian Conflict* 14–26.

115. Ibid., 15.

116. Ibid., 7–8, 24.

117. Ibid., 25. Cf. Lloyd, *Magic, Reason and Experience* 49–58 (on the Hippocratic Corpus).

118. Remus, *Pagan-Christian Conflict* 182. For a case study of a highly rational and cultivated second-century Greek, see B. S. MacKay, "Plutarch and the Miraculous," in *Miracles: Cambridge Studies in Their Philosophy and History*, ed. C. F. D. Moule, 95–111.

119. This has been denied, but see the judicious conclusions of Newmyer, "Talmudic Medicine and Greco-Roman Science" 2904. The extent to which Jewish physicians were influenced by Greek medical theories is debated, but Talmudic medicine shares with Greek medicine a reliance on the rational observation of nature (ibid., 2902 and 2903).

120. See Lk. 13:1–5 and Jn. 9:1–4 (cf. 11:4).

121. Κύριε, εἰ κεκοίμηται σωθήσεται.

122. For example, Everett Ferguson believes that the "the spirit that had crippled her" (Lk. 13:11) "suggests a physical ailment caused by a spirit; thus this is a healing miracle and not an expulsion or exorcism of an evil spirit" (*Demonology of the Early Christian World* 12). His view is typical of those scholars who do not consider the condition to be one of possession.

123. "Nevertheless, it may be said that [in the New Testament] the existence of sickness in the world belongs to the character of the αἰὼν οὗτος of which Satan is the prince" (*TDNT* 2:18, s.v. δαίμων κτλ [Foerster]).

124. See esp. vv. 11 (πνεῦμα ἔχουσα ἀσθενείας) and 16 (ἣν ἔδησεν ὁ ατανᾶς).

125. ὃς διῆλθεν εὐεργετῶν καὶ ἰώμενος πάντας τοὺς καταδυναστευομένοις ὑπὸ τοῦ διαφόλε.

126. I take the phrase "a spirit that had crippled her" in Luke 13:11 to refer to the activity of Satan as the source of evil. Not only is there no hint of possession, but the language does not seem to imply a proximate demonic causation (contra Aune, "Magic in Early Christianity" 1529 n. 97). Jesus frees the woman not from a spirit but from her physical infirmity (Γύναι, ἀπολέλυσαι τῆς ἀσθενείας σου [v. 12]), and by laying his hands on her in a symbolic

gesture of healing he restores her deformed posture (v. 13). The words of Foerster are apposite here: "No balance or clear-cut distinction is attempted between natural and Satanic elements; the 'murderer from the very beginning' is secretly behind the phenomenon of sickness" (*TDNT* 7:159, s.v. σατανᾶς).

127. See Amundsen and Ferngren, "Perception of Disease" 2952–5.

128. J. B. Cortés and F. M. Gatti, *The Case against Possessions and Exorcisms: A Historical, Biblical, and Psychological Analysis of Demons, Devils, and Demoniacs* 103.

129. For an attempt to establish a causal relationship, see Kelly, *The Devil, Demonology, and Witchcraft* 70–1. Kelly believes that "in the gospel of Luke the connection between disease, demons, and the devil is strikingly presented." He cites four episodes in Luke as illustrating this connection: two in Lk. 4:33–39, in which Jesus rebukes both a demon (v. 35) and the fever of Peter's mother-in-law (v. 39); Lk. 10:9–20, in which the seventy return rejoicing that even the demons are subject to them; and Lk. 13:11–16, in which Jesus heals the woman afflicted with the spirit of infirmity (ibid., 71). Of the four episodes the first is the most promising for his thesis, since Jesus is said to have "rebuked" spirits on other occasions (see, e.g., Mk. 9:25). The same verb (ἐπιτιμᾶν), however, is used not only of Jesus's exorcising demons but of his bending the inanimate forces of nature to his will (e.g., in Lk. 8:24, where he tames the winds and the sea in a passage that features no hint of demonic agency). Since the fever (πυρετός) of Peter's mother-in-law is not attributed to demonic agency in Luke's narrative, the only basis for assuming a demonic etiology is the antecedent belief that the Gospels ascribe all disease to demons. Luke 4:39 should probably be taken rather as a case of Jesus's subjecting disease to his authority (cf. *TDNT* 2:626, s.v. ἐπιτιμάω; contra Aune, "Magic in Early Christianity" 1530–1).

130. "How many things there are in us which, if they persist, bar our entry into the kingdom of heaven! Just think, brothers and sisters, how urgently people beg doctors for merely temporary health, how if someone is desperately ill he's neither slow nor shy about clinging to the man's feet, about washing the expert surgeon's feet with his tears" (Sermon 80.3, in Augustine, *Sermons,* in Rotelle, *Works of St. Augustine* 3:352). Cf. Sermons 84.1 and 344.5.

131. Larry Hogan calls this "the most important statement about healing and sickness in the Hebrew Scriptures" (L. P. Hogan, *Healing in the Second Temple Period* 4).

132. Sir. 38:1, 6–7, 9 (New Revised Standard Version modified).

133. Ambrose, *On Cain* 1.40. The theme is a common one. Cf. Jerome, *On Isaiah* 8, and Augustine, Tractate 30 on John 3; *On Christian Doctrine* 4.16.33; Sermon 84; for Basil, see E. F. Morrison, *St. Basil and His Rule: A Study in Early Monasticism* 127.

134. Lane Fox, *Pagans and Christians* 328.

135. See C. Herzlich and J. Pierret, *Illness and Self in Society* 98–125, esp. 98–101; on recourse to parallel therapies, 206–9.

136. Cf. R. Van Dam, *Saints and Their Miracles in Late Antique Gaul* 115.

137. Merideth 30–33; pace Peter Brown, who writes that later generations, unlike earlier Greeks and Romans, were more concerned with protecting themselves from demonic attacks than diagnosing "the intimate disorders of their constitutions" (*The World of Late Antiquity* 56).

138. Cf. Martin, *The Corinthian Body* 152.

139. Vivian Nutton contrasts Greek with Christian explanations of disease (he assumes that early Christians believed in demonic etiology): "Individual illness is thus less caught up in the eternal struggle between the mysterious forces of good and evil; bodily infirmity is explicable on physical not moral grounds. In these circumstances [i.e., on the Greek understanding], religion and medicine coexist, even co-operate" ("Murders and Miracles" VIII 48–9). Christians did not believe, howevever, that moral failings were directly the cause of sickness but rather that God sometimes brought physical suffering upon believers as a means of spiritual discipline. They did not thereby deny the operation of natural causes. One finds the simultaneous acceptance of both in many of the church fathers, as, for example, in the Cappadocians.

Chapter 4 • Christianity as a Religion of Healing

1. On the place of medicine and healing in the early church, see A. Harnack, "Medicinisches aus der ältesten Kirchengeschichte," in *Texte Untersuchungen zur Geschichte der altchristlichen Literatur* vol. 8, pt. 4, 37–152; Frings, "Medizin und Arzt bei den griechischen Kirchenvätern bis Chrysostomos"; H. Schadewaldt, "Die Apologie der Heilkunst bei den Kirchenvätern," *Veröffentlichungen der internationalen Gesellschaft für Geschichte der Pharmazie* 26 (1965): 115–30; Amundsen, "Medicine and Faith," in Amundsen 127–57 and 5–12; Temkin, *Hippocrates in a World of Pagans and Christians;* Dörnemann, *Krankheit und Heilung* 182–5, 288–98.

2. On Christianity as a religion of healing, see Harnack, *The Mission and Expansion of Christianity* 1:121–51. Harnack's views have been widely adopted: see, e.g., Case, "The Art of Healing" 253–5; S. Angus, *The Religious Quests of the Graeco-Roman World: A Study in the Historical Background of Early Christianity* 414–38; Edelstein and Edelstein, *Ascelpius* 2:133–4 and n. 4; and Brown, *The World of Late Antiquity.*

3. A. Harnack, "The Gospel of the Saviour and of Salvation," in *Mission and Expansion of Christianity* 1:121–51, develops this theme; quotation at 131–2.

4. Case, "Art of Healing" 253; cf. R. MacMullen: "The chief business of religion . . . was to make the sick well" (*Paganism in the Roman Empire* 49). Case discusses the healing function of ancient religions (which he overemphasizes) in *Experience with the Supernatural in Early Christian Times* 221–63.

5. Nutton, "From Galen to Alexander" X 5.

6. Ibid. Cf. Nutton, "Murders and Miracles" VIII 45–51. "By contrast [with paganism]," writes Nutton, "from its inception Christianity offered itself as a direct competitor to secular healing" (48). In a private communication Professor Nutton tells me that he has modified this position. He writes, "This type of Christian healing is always a minority position and, even when adopted, does not *always* accompany a total rejection of secular healers."

7. On the allegedly competitive nature of Christian miraculous healing see Brown, *The Cult of the Saints* 118, 120.

8. The literature on the miracles of Jesus is extensive and almost unmanageable. For a comprehensive treatment, see H. van der Loos, *The Miracles of Jesus*, trans. T. S. Preston; also A. Richardson, *The Miracle Stories of the Gospels.* On the method of approach to New Testament miracles, see H. C. Kee, *Miracle in the Early Christian World: A Study in Socio-*

historical Method, esp. 146–73 and 290–6; and, on the healing miracles in particular, Kee, *Medicine, Miracle, and Magic in New Testament Times* 75–9. Kee offers a necessary corrective to the form-critical approach of Seybold and Mueller (*Sickness and Healing* 114–29), but for a methodological critique, see Amundsen and Ferngren, "The Healing Miracles of the Gospels." For a discussion of the various ways in which Jesus's role as miracle worker has been viewed (e.g., physician, divine man, magician, suggestive healer, shaman), see B. Kollmann, *Jesus und die Christen als Wundertäter. Studien zu Magie, Medizin und Schamanismus in Antike und Christentum* 31–42.

9. Peter's mother-in-law (Mk. 1:29–31 = Mt. 8:14–15 = Lk. 4:38–39); a leper (Mk. 1:40–45 = Mt. 8:1–4 = Lk. 5:12–16); a paralytic (Mk. 2:1–12 = Mt. 9:1–8 = Lk. 5:17–26); a man with a withered hand (Mk. 3:1–6 = Mt. 12:9–14 = Lk. 6:6–11); the daughter of Jairus (Mk. 5:21–43 = Mt. 9:18–26 = Lk. 8:40–56); a woman with a hemorrhage (Mk. 5:25–34 = Mt. 9:20–22 = Lk. 8:43–48); a deaf mute (Mk. 7:31–36); the blind man at the pool of Bethsaida (Mk. 8:22–26); blind Bartimaeus (Mk. 10:46–52 = Mt. 9:27–34 = Lk. 18:35–43); a young man at Nain (Lk. 7:11–17); a deformed woman (Lk. 13:10–17); ten lepers (Lk. 17:11–19); a man with dropsy (Lk. 14:1–6); a paralytic (Jn. 5:1–9); the raising of Lazarus (Jn. 11); a man born blind (Jn. 9:1–41); the centurion's servant (Mt. 8:5–13 = Jn. 4:46–54) (Aune, "Magic in Early Christianity" 1523–24 n. 68, whose list I reproduce here).

10. Ladd, *The Presence of the Future* 211–13.

11. See Brown, *The New International Dictionary of New Testament Theology* 2: 626–35, s.v. "Miracle"; C. F. D. Moule, "The Vocabulary of Miracle," in *Miracles* 235–8.

12. E. Wright and R. H. Fuller, *The Book of the Acts of God: Contemporary Scholarship Interprets the Bible* 272–5.

13. See J. Wilkinson, "A Study of Healing in the Gospel According to John," *SJT* 20 (1965): 457–60. As Vivian Nutton reminds me, some readers of the Gospels might not have taken Jesus's healings in a purely soteriological sense. Moreover, the fact that Gospels attribute miracles of healing only to Jesus and his disciples does not exclude the possibility that some Christians might claim them as well.

14. *TDNT* 1:433–34, s.v. ἀπόστολος.

15. See Hardon, "The Miracle Narratives in the Acts of the Apostles" 303–5.

16. Case, "Art of Healing" 254.

17. Pace Case, who misinterprets Peter's address in his attempt to demonstrate that early Christianity was a religion of healing ("Art of Healing" 255). The apostles mentioned healing only in a general way in their preaching, and then with reference primarily to Jesus (as, e.g., in Acts 10:38).

18. On the contrast between salvation in the New Testament and the concept in later Judaism, Greek thought, and Gnosticism, see *TDNT* 7: 1002–3, s.v. σώζω.

19. Cf. Aune, "Magic in Early Christianity" 1548.

20. On the date of James, see Guthrie, *New Testament Introduction* 761–4; R. M. Grant, *A Historical Introduction to the New Testament* 222.

21. See M. Goguel, *The Primitive Church* 371–7; J. Wilkinson, "Healing in the Epistle of James," *SJT* 24 (1971): 326–45; J. B. Mayor, *The Epistle of St. James: The Greek Text with Introduction*, 3d ed., 169–74; and Kollmann, *Jesus und die Christen als Wundertäter* 344–7. On the anointing of the sick, see *TDNT* 1:230–2, s.v. ἀλείφω.

22. Morton Smith cites the following references to illness in the New Testament Epistles: 1 Cor. 2:3, 11:30; 2 Cor. 4:10 ff., 11:30, 12.5, 7–10; Gal. 4:13 ff.; Phil. 2:26 ff., 3:10; Col. 1:24; 1 Tim. 5:23; 2 Tim. 4:20 ("De tuenda sanitate praecepta" 40 n. 24). It is unlikely, however, that physical illness is referred to in every case.

23. On the nature of Paul's "thorn" (*skolops*), which is usually assumed to be a physical affliction, see Seybold and Mueller, *Sickness and Healing* 171–82, and R. P. Martin, *2 Corinthians, Word Biblical Commentary* 40:410–16. Not all commentators have been convinced that his ailment was a physical one, but it remains the most common view. An enormous literature has arisen in which (by one count) at least twelve suggestions have been offered, including a pain in the head or ear, ophthalmia (cf. Gal. 4:13–15), epilepsy, neuralgia, defective speech, and depression.

24. Arndt and Gingrich, *A Greek-English Lexicon* 403, s.v. κύμνω (though the verb in James 5:15 is taken, I believe mistakenly, to refer to physical illness); *TDNT* 7:990, s.v. σώζω.

25. See, e.g., 1 Cor. 1:21; Heb. 1:9; 1 Jn. 2:20, 27; Rev. 3:18. One might infer that the prayer of the presbyters in James itself constitutes the anointing.

26. On Montanism, see below, pp. 71–72.

27. Tertullian, *Ad Scapulam* 4.

28. Vocet presbyteros ecclesiae et imponant ei manus unguentes eum oleo (Homily 2 on Leviticus 4). Goguel takes the imposition of hands as evidence that Origen regarded the text as prescribing a rite of reconciling penitents who sought forgiveness of their sins (*The Primitive Church* 375).

29. John Chrysostom, *On the Priesthood* 3.6.

30. Origen makes no reference to healing. Chrysostom quotes the passage in James in the context of a discussion of the power of the presbyter to forgive sins, and he mentions physical healing only incidentally.

31. See *Canons of Hippolytus* 21C. The prayer of anointing (prayer 17) in the *Sacramentary of Serapion* (quoted in Merideth 103) is too general to be considered a specific rite of healing from sickness. See also C. W. Gusmer, *And You Visited Me: A Sacramental Ministry to the Sick and the Dying* 5–21, and J. L. Empereur, *Prophetic Anointing: God's Call to the Sick, the Elderly, and the Dying* 121–37.

32. F. S. Paxton, *Christianizing Death: The Creation of a Ritual Process in Early Medieval Europe* 27–32; 55–9; 70–3.

33. Warfield 3–31; but see n. 35 below. For what follows I am indebted to Warfield's discussion; see also Dörnemann, *Krankheit und Heilung* 80–7.

34. On the lack of miraculous elements generally in the Apostolic Fathers, see Warfield 10–11; cf. J. B. Lightfoot, ed. and trans., *The Apostolic Fathers, pt. 2, S. Ignatius, S. Polycarp* 1: 614–16.

35. Achtemeier, "Jesus and the Disciples as Miracle Workers" 156–61.

36. Justin Martyr, *Second Apology* 6.

37. Irenaeus, *Against Heresies* 2.32.4; cf. 2.31.2.

38. Origen states that "traces" of prophecy and miracles are still to be found, and he claims to have witnessed them (*Contra Celsum* 2.8) but does not cite specific examples.

39. Warfield 11–16; cf. J. S. McEwen, "The Ministry of Healing," *SJT* 7 (1954): 136–41.

40. Irenaeus regards the ability to perform miracles, especially of healing and raising the

dead, as an authenticating sign of orthodox Christianity that heretics lack. For a survey of controversies regarding the place of miracles in early Christian thought, see M. van Uytfanghe, "La Controverse biblique et patristique autour du miracle, et ses répercussions sur l'hagiographie dans l'Antiquité tardive et le haute Moyen Âge latin," in *Hagiographie, cultures et sociétés IV–XII siécles* 205–33.

41. Origen, *Contra Celsum* 1.46. On this passage, see Warfield 239–40 n. 22.

42. Tertullian, *De pudicitia* 21. See Warfield 14–15. Tertullian mentions several instances of exorcism but only one of healing (*Ad Scapulam* 4).

43. McEwen, "The Ministry of Healing" 137–8.

44. The tendency of second-century writers like Herodes Atticus, Marcus Aurelius, and others to dwell on physical ills has been much commented upon (see, e.g., Dodds, *Pagan and Christian* 39–45; Bowersock, *Greek Sophists* 71–3, on the correspondence of Fronto and Marcus Aurelius; and J. E. G. Whitehorne, "Was Marcus Aurelius a Hypochondriac?" *Latomus* 36 [1977]: 413–21, also on the correspondence).

45. Edelstein and Edelstein, *Asclepius* 2:255.

46. On Isis and Serapis, see R. E. Witt, *Isis in the Graeco-Roman World* 185–97, and Kee, *Miracle in the Early Christian World* 105–45.

47. On Apollonius, see M. Dzielska, *Apollonius of Tyana in Legend and History,* trans. P. Pieńkowski; as the "pagan Christ," see 15. On charismatic healers generally, see Smith, "Prolegomena," and J. Z. Smith, "Good News Is No News: Aretalogy and Gospel," in *Christianity, Judaism and Other Greco-Roman Cults: Studies for Morton Smith at Sixty,* ed. J. Neusner, 21–38.

48. R. MacMullen, *Paganism in the Roman Empire* 95–6, 135; for a critique, see Praet, "Explaining the Christianization of the Roman Empire" 9–11, 25–7.

49. MacMullen, *Christianizing the Roman Empire* 27; cf. idem, "Two Types of Conversion to Early Christianity," in *Changes in the Roman Empire: Essays in the Ordinary* 140–1; see also Harnack, *Mission and Expansion* 1:131; Brown, *The World of Late Antiquity* 55 and "The Rise and Function of the Holy Man in Late Antiquity," in *Society and the Holy in Late Antiquity* 121–6, originally published in *Journal of Roman Studies* 61 (1971): 80–101; and Perkins, *The Suffering Self* 17–18.

50. MacMullen, *Paganism in the Roman Empire* 50.

51. Leeper, "Exorcism in Early Christianity" 167.

52. See Edelstein and Edelstein, *Asclepius* 2:132–3.

53. M. F. Wiles, "Miracles in the Early Church," in Moule, *Miracles* 222.

54. *Haer.* 2:48–9.

55. Lampe, "Miracles and Early Christian Apologetic," in Moule, *Miracles* 215. On the widespread belief that miracles had declined in the second and third centuries, see Uytfanghe, "La Controverse biblique et patristique" 210–11.

56. *Contra Celsum* 1.46.

57. Lampe, "Miracles and Early Christian Apologetic" 215.

58. Lane Fox, *Pagans and Christians* 327–30, quotation at 330.

59. Ibid., 328, 329.

60. Origen, *Contra Celsum* 3.24–5; on this passage, see Angus, *Religious Quests of the Graeco-Roman World* 420–2. On Celsus generally, see R. L. Wilken, *The Christians As the*

Romans Saw Them 94–125. On healing by malevolent forces (pagan gods and demons), see Amundsen 7.

61. "The [old] faith thus practically engendered and rewarded proved a stubborn barrier to the onward march of Christianity. Paganism made its last stand in the temples of Serapis and Aklepios, and their powers of resistance were due to the cures performed under their auspices in the name of the god" (Hamilton, *Incubation* 109).

62. Lane Fox, *Pagans and Christian* 330–1.

63. The chief sources for Montanism are Eusebius, *Eccles. Hist.* 5.3.4, 5.14–18, and Epiphanius's *Panarion* (*Haeresis*) 48. See F. C. Klawater, "The New Prophecy in Early Christianity: The Origin, Nature and Development of Montanism, A.D. 165–220"; D. E. Aune, *Prophecy in Early Christianity and the Ancient Mediterranean World;* and C. Trevett, *Montanism: Gender, Authority and the New Prophecy.*

64. Eusebius dates the origin of the sect to 172 (*Chronicle*), while Epiphanius places it somewhat earlier, c. 156–57 (*Panarion* 48.1.1). The latter date appears more likely: see T. D. Barnes, "The Chronology of Montanism," *Journal of Theological Studies,* n.s., 20 (1970): 403–408.

65. *De anima* 9.4. On Montanism in North Africa, see Barnes, *Tertullian* 130–42.

66. Lane Fox, *Pagans and Christians* 410 and 748 n. 27.

67. Lampe, "Miracles and Early Christian Apologetic" 215.

68. Epiphanius relates that one of the early Montanist prophetesses (either Quintilla or Priscilla) was sleeping in Pepuza when, in a dream, Christ appeared to her in the form of a woman and revealed that Pepuza was a holy site and that it was there that Jerusalem would come down from heaven. "For this reason they say that even today certain women and men are initiated in the same way there in that place [i.e., by means of incubation?] so that the women and men who wait there may see Christ" (*Panarion* 490.2–4).

69. S. Gero, "Montanus and Montanism according to a Medieval Syriac Source," *Journal of Theological Studies,* n.s., 28 (1977): 520–4. Attempts to account for Montanism's peculiar practices as having been borrowed from pagan cults in Asia Minor have not met with wide acceptance. "All of the major features of early Montanism, including the behavior associated with possession trance, are derived from early Christianity" (Aune, *Prophecy in Early Christianity* 313).

70. Irenaeus, *Against Heresies* 2.31.2.

71. A convenient selection of these works with good introductions is that of E. Hennecke, *New Testament Apocrypha,* ed. W. Schneemelcher with English translation by R. McL. Wilson, 2 vols. See also J. Hastings, ed., *A Dictionary of Christ and the Gospels* 1: 671–85, s.v. "Gospels (Apocryphal)."

72. Achtemeier, "Jesus and the Disciples as Miracle Workers" 161.

73. Ibid., 162–73.

74. On healing in the early Christian apocryphal works, see Kee, *Miracle in the Early Christian World* 274–89 and Dörnemann, *Krankheit und Heilung* 69–79. Kee points out that ancient romances, both pagan and Christian, often "served as a vehicle for conveying religious truth or as an apology for a philosophical view" (252). Some of the apocryphal gospels seek to foster ascetic practices or the denigration of the body, while others probably reflect the desire for a direct personal miraculous experience. Though arising from a heterodox milieu, they constitute the first examples of Christian aretalogies.

75. Achtemeier, "Jesus and the Disciples as Miracle Workers" 164–5.

76. Ibid., 172–3.

77. Perkins, *The Suffering Self* 173–99.

78. Ibid., 9. See Reff, *Plagues, Priests, and Demons* 9–10 for a critique.

79. Perkins, *The Suffering Self* 124–41.

80. For a detailed introduction to the Acts of Peter, bibliography, and English translation by W. Schneemelcher, see Hennecke, *New Testament Apocrypha* 2:259–322.

81. Perkins, *The Suffering Self* 129.

82. Ibid. On its theological tendency, see Hennecke, *New Testament Apocrypha* 2:274–5.

83. Cameron, *Christianity and the Rhetoric of Empire* 89–119, quotation at 117.

84. G. W. H. Lampe, "Miracles in the Acts of the Apostles," in Moule, *Miracles* 165–6.

85. The point is made by Merideth 73–74.

86. For a survey of the literature, see Leeper, "Exorcism in Early Christianity" 160–2.

87. Schneemelcher cites Schmidt, who thinks that the work long enjoyed popularity in Catholic circles (Hennecke, *New Testament Apocrypha* 268) in spite of its Encratite ethics and Docetic Christology. It was as a piece of popular literature, "more concerned with edification and moral effect," that it attracted readers (ibid., 275).

88. Lampe, "Miracles and Early Christian Apologetic" 206.

89. Leeper, "Exorcism in Early Christianity" 161–2.

90. "Although the legendary 'acts' of Apostles laid great weight on the signs and wonders which their heroes worked, they were not historical texts, nor were they written to win pagan converts: they aimed to impress Christian readers and spread the views of a minority of fellow Christians through vivid fiction" (Lane Fox, *Pagans and Christians* 329). Yet Vivian Nutton observes in a private communication that "a side stream is still a part of the river," i.e., that the writers of the apocryphal acts represented the views of one group of professed Christians, even if it lay outside the mainstream.

91. Ibid., 330. William Babcock points out that the distinction between elites and the masses in early Christianity was not firm. It is not unusual to find "high brow" and "low brow" elements (e.g., philosophical theology and popular religion) in the same persons (W. S. Babcock, "MacMullen on Conversion: A Response," *Second Century* 5 [1985/86]: 82–89).

92. Gregory of Nyssa, *De deitate Filii et Spiritus Sancti (Patrologia Graeca* 46:557), as quoted in P. Brown, *Power and Persuasion in Late Antiquity: Towards a Christian Empire* 89–90.

93. Lane Fox, *Pagans and Christians* 330.

94. Quoted by Reff, *Plagues, Priests, and Demons* 65.

95. Lane Fox, *Pagans and Christians* 330–1.

96. See Cyprian, *Quod idola dii non sint* 7, and perhaps his Epistle 75.15 (although I am not certain that Cyprian has in mind here physical healing). It is important to note in this context that Cyprian was disappointed by the fact that many of his congregants were terrified of death and believed that in time of plague they would be spared because they were Christians. On the metaphorical use of medicine in Cyprian, see Dörnemann, *Krankheit und Heilung* 172–9.

97. This was Gibbon's view (*The History of the Decline and Fall of the Roman Empire,* ed. J. B. Bury, 3:225–7 [bk. 2, ch. 28, 4]; see Greer, *The Fear of Freedom* 117–18). See also Hamilton,

Incubation 109–71, on the borrowing of pagan healing practices by Christians, and Csepregi, "The Compositional History of Greek Christian Incubation Miracle Collections."

98. On the respective views of E. R. Dodds and Peter Brown regarding the "crisis of the third century," see J. G. Gager, "Introduction: The Dodds Hypothesis," in *Pagan and Christian Anxiety: A Response to E. R. Dodds*, ed. R. C. Smith and J. Lounibos, 1–11, and P. Brown, "Approaches to the Religious Crisis of the Third Century A.D.," in *Religion and Society in the Age of Saint Augustine* 74–81 (originally published in *English Historical Review* 83 [1968]: 542–58).

99. It has been effectively challenged by several scholars, including Gager (see previous note) and Ramsay MacMullen (see the latter's *The Roman Government's Response to Crisis, A.D. 235–337* 13–16); see also Praet, "Explaining the Christianization of the Roman Empire" 68–70. MacMullen observes that many of the features of the alleged pessimism and mysticism of the third century could equally well be found in any era at all.

100. Brown, "The Rise and Function of the Holy Man in Late Antiquity" 148; idem., *The Cult of the Saints* 13–22; and idem, *Authority and the Sacred* 57–78, 85–7. On Brown's study of the holy man, see Howard-Johnston and Hayward, *The Cult of the Saints in Late Antiquity and the Middle Ages.*

101. See Rom. 8:13; 1 Cor. 9:24–27, Col. 3:5.

102. See, e.g., Clement of Alexandria, *Stromateis* 3.7; Lactantius, *The Divine Institutes* 5.22; Ambrose, Epistle 63.91; Augustine, *Of True Religion* 20.40. See also Amundsen, "Medicine and Faith in Early Christianity," reprinted in Amundsen 134; F. Bottomley, *Attitudes to the Body in Western Christendom*; and M. Miles, *Fullness of Life: Historical Foundations for a New Asceticism.*

103. See Dodds, *Pagan and Christian* 7–36.

104. See Frend, *Martyrdom and Persecution* 356, 548.

105. See F. van der Meer, *Augustine the Bishop: The Life and Work of a Father of the Church,* trans. B. Bettershaw and G. R. Lamb, 527–57. Crislip observes that "religious healing was not just a product of the hagiographic imagination. Private letters from Egypt, dated to the 340s and 350s, confirm the roles of monastics as religious healers, especially to nonmonastic followers" (Crislip 22). They support the close association of religious healing with ascetics without undercutting Greer's broader thesis of a change in sensibility in the fourth century.

106. Credulity was not limited to Christian intellectuals but was characteristic of pagan intellectuals as well (see Nilsson, *Greek Piety* 165–6). It was a feature of the spirit of the age.

107. Warfield 37–38.

108. Augustine, *Of True Religion* 25.47 (on this passage and Augustine's changing views of miracles see Greer, *The Fear of Freedom* 170–8); cf. his Sermones 88.2.3. In his *Retractations,* which he wrote at the end of his life, he modified his position in *Of True Religion.*

109. P. Brown, *Augustine of Hippo: A Biography* 414. Donatism was a schism that divided the church in North Africa during the fourth and early fifth centuries.

110. In fact, as early as 415 he had begun to change his mind about the genuineness of reported miracles: see van der Meer, *Augustine the Bishop* 540.

111. Augustine, *City of God* 22.8; Brown, *Augustine of Hippo* 416; cf. A. Momigliano, "Popular Religious Beliefs and the Late Roman Historians," in *Popular Belief and Practice*, ed. G. J. Cuming and D. Baker, 17–18.

112. Greer, *The Fear of Freedom* 33–4.

113. Ibid., 90–1.

114. Ibid., 86–7.

115. Ibid., 46–7.

116. Ibid., 48.

117. Ibid., 49.

118. Ibid., 81.

119. Ibid., 5.

120. Ibid., 92, 115–6. But Augustine demonstrates that varying attitudes (e.g., providentialism) were exhibited in the fifth century.

121. See A. A. Barb, "The Survival of Magic Arts," in *The Conflict between Paganism and Christianity in the Fourth Century*, ed. A. Momigliano, 115; Dodds, *Pagan and Christian* 125–6; A. H. M. Jones, *The Later Roman Empire, 284–602: A Social, Economic, and Administrative Survey* 2:962; Brown, *The Making of Late Antiquity* 60–6. For a survey of the practice of magic in the Roman Empire, see R. MacMullen, *Enemies of the Roman Order: Treason, Unrest and Alienation in the Empire* 95–127.

122. Magic in the Roman world carried the penalty of deportation or death. See C. Pharr, "The Interdiction of Magic in Roman Law," *TAPA* 63 (1932): 269–95.

123. *Codex Theodosianus* 9.16.3.

124. See Augustine, *City of God* 10.9. On Augustine's views of magic generally (and those of Caesarius of Arles and Isidore of Seville), see M. Bailey, *Magic and Superstition in Europe: A Concise History from Antiquity to the Present* 53–59. See also B. Ward, *Miracles and the Medieval Mind: Theory, Record, and Event, 1000–1215*, rev. ed.; and N. Janowitz, *Magic in the Roman World: Pagans, Jews and Christians* 16–20.

125. "There have always been amulets, but belief in and use of them increased under the Empire in an unheard-of manner" (Nilsson, *Greek Piety* 167). This was true of Christians as well as pagans (see D. Frankfurter, "Amuletic Invocations of Christ for Health and Fortune," in *Religions of Late Antiquity in Practice*, ed. R. Valantasis, 340–3). But Christian writers, decrees of councils, and Roman legislation beginning with Constantine almost universally condemned them (Bailey, *Magic and Superstition* 50–53).

126. John Chrysostom, Homily 8 on Colossians.

127. On the softening of Christian attitudes, see V. I. J. Flint, *The Rise of Magic in Early Medieval Europe* 29–35.

128. R. MacMullen, "Constantine and the Miraculous," in *Changes in the Roman Empire: Essays in the Ordinary* 111; Greer, *The Fear of Freedom* 119–20.

129. See M. E. Keenan, "Augustine and the Medical Profession," *TAPA* 67 (1936): 184.

130. Ibid.

131. Augustine, *On Christian Doctrine* 2.45.

132. See C. Jenkins, "Saint Augustine and Magic," in *Science, Medicine, and History: Essays on the Evolution of Scientific Thought and Medical Practice Written in Honour of Charles Singer*, ed. E. A. Underwood, 1:135, and Keenan, "Augustine and the Medical Profession" 184.

133. Augustine, *Confessions* 1.10.17. Augustine writes that his mother would have allowed him to be baptized had he been about to die but she feared that if he lived his future sins would bring even greater guilt.

134. Greer, *The Fear of Freedom* 123.

135. See B. J. Cooke, *Ministry to Word and Sacraments: History and Theology* 356; Jones, *Later Roman Empire* 2:962–3.

136. Crislip 22.

137. *DCA* 2:2042, s.v. "Wonders."

138. Gregory Thaumaturgus, one of the most prominent miracle workers, became bishop of Neocaesarea in Pontus, Asia Minor. He was celebrated for the many miracles (including healing miracles) that were attributed to him, accounts of which are preserved in three or four lives that were written independently in Latin, Syriac, and Armenian. Raymond van Dam argues that they are based on oral tradition in which the stories of his wonder-working have symbolic value. They draw on common literary themes from folklore and on legendary elements that are characteristic of hagiographic literature ("Hagiography and History: The Life of Gregory Thaumaturgus," *Classical Antiquity* 1 [1982]: 280, 284, 286). It is possible that some of the accounts of Gregory date from his own lifetime. But I would not argue that there exists *no* cases in which popular tradition ascribed miracle-working powers to some individuals in the first four centuries of Christianity. Indeed, one would expect that to occur in any religious tradition. But as compared with the fifth century and later, these individuals were rare among Christians, even in such sectarian traditions as Montanism. On Gregory of Nyssa's panegyric of Gregory Thaumaturgus, which distorts his own description of his early life, see Lane Fox, *Pagans and Christians* 528–39.

139. On the magical use of Jesus's name, see Aune, "Magic in Early Christianity" 1545–8.

140. On the development of the cult of martyrs, see van der Meer, *Augustine the Bishop* 471–97.

141. See Nutton, "From Galen to Alexander" X 7.

142. Jones, *Later Roman Empire* 2:961.

143. Greer, *The Fear of Freedom* 134–5.

144. Ibid., 148–9.

145. Ibid., 180–1.

146. MacMullen, "Constantine and the Miraculous" 107–16, 312–6.

147. Ibid., 114.

148. Cf. P. Brown, "Sorcery, Demons, and the Rise of Christianity from Late Antiquity into the Middle Ages," in *Witchcraft Confessions and Accusations*, ed. M. Douglas, 28–29, 31–3; reprinted in P. Brown, *Religion and Society in the Age of Saint Augustine* 119–46.

149. Remus, *Pagan-Christian Conflict* 89–90, 193–4.

150. See Finn 137–47. Though Finn is dealing with the encouragement of almsgiving in sermonic literature, his words have a wider application; cf. also Merideth 152–81, who focuses on the rise of a "discourse of disease" that was widely influential, and Leeper, "Exorcism in Early Christianity" 162–4.

151. See MacMullen, "Constantine and the Miraculous" 112; see above, pp. 69–70.

152. See Nock's discussion of the factors that led to the conversions of Justin Martyr, Arnobius, and Augustine (*Conversion* 254–71).

153. Dodds quotes Nock's description of the cult of Asclepius as "a religion of emergencies" (Dodds, *The Greeks and the Irrational* 203 n. 83).

154. Merideth 29.

155. Browning, "The 'Low Level' Saint's Life in the Early Byzantine World" 121–2.
156. *DCA* 2:2042–4, s.v. "Wonders."
157. See Augustine, *City of God* 8.160; cf. 8.350–3.

Chapter 5 • The Basis of Christian Medical Philanthropy

1. F. Frend, *Martyrdom and Persecution in the Early Church* 176–7. For a survey of modern literature on the persecution of the early Christians, see Praet, "Explaining the Christianization of the Roman Empire" 29–33. Aline Rousselle observes that persecution was neither widespread nor continuous before the reign of Diocletian, when it occasioned real atrocities against Christians. Yet even in times of relative peace Christians lived in fear of torture and death (see A. Rousselle, *Porneia: On Desire and the Body in Antiquity* 130–1 and the literature cited in nn. 2 and 3).

2. By the end of the first century there existed some forty or fifty cities in the Roman Empire that had a Christian congregation; the total number of Christians was "probably less than fifty thousand" (Reff, *Plagues, Priests, and Demons* 65 and nn. 139 and 140 for citations of the secondary literature from which he derives these numbers). For a sociological attempt to provide quantitative analysis of the growth of urban Christianity, see Stark 129–45. Stark calculates that there were about 1,000 Christians in A.D. 40; 7,530 by the end of the first century; 217,795 by the end of the second; 6.3 million by the end of the third; and more than 33 million by the middle of the fourth (7). He projects a growth rate of 3.4 percent each year. See also K. Hopkins, "Christian Number and Its Implications," *Journal of Early Christian Studies* 6 (1998): 185–226.

3. On *philanthropia,* see J. Ferguson, *Moral Values in the Ancient World* 102–17, and R. le Déaut, "Philanthropia dans la littérature grecque Jusqu'au Nouveau Testament," in *Studi e Testi: Mélanges Eugène Tisserant* 1:255–94.

4. *Lives of Eminent Philosophers* 3.98.

5. R. Garrison, *Redemptive Almsgiving in Early Christianity* 38–45.

6. A. R. Hands, *Charities and Social Aid in Greece and Rome* 77–88; G. Downey, "Who Is My Neighbor? The Greek and Roman Answer," *Anglican Theological Review* 47 (1965): 2–15; H. Fashar and J. Jouanna, eds., *Médicine et morale dans l'antiquité.* For a survey of scholarship on the history of Christian charity, see Finn 26–33. Approaches to the study of early Christian charity generally reflect one of three ideological perspectives: those that emphasize evolution, viewing Christian charity as an improvement on Graeco-Roman philanthropy; those that emphasize continuity, considering Christian practices merely a continuation of Graeco-Roman philanthropy; and those that emphasize civic identity, exploring the cultural context of early Christian charity, such as patronage or euergetism (Holman, *The Hungry Are Dying* 6–12). On euergetism (an economy based on gift giving), which dominated the classical world for a thousand years, see P. Veyne, *Bread and Circuses: Historical Sociology and Political Pluralism,* trans. B. Pearce. I accept the view of H. Bolkestein (*Wohltätigkeit und Armenpflege in Vorchristlichen Altertum*), Hands (*Charities and Social Aid*), and Veyne that Christian charity represents a radical new departure from classical euergetism, a euergetism that was grounded in religion (see *Bread and Circuses* 19–34, esp. 19). Richard Finn modifies this view

in arguing that classical and Christian patterns of philanthropy had previously been characteristic of secular rulers (Finn 208–14, 218–20; so also Brown, *Poverty and Leadership* 1–44).

7. See *TDNT* 6:39–40, s.v. πένης κτλ, and 6:888–902, s.v. πτωχός; N. W. Porteous, "The Care of the Poor in the Old Testament," in *Service to Christ: Essays Presented to Karl Barth on his 80th Birthday,* ed. J. I. McCord and T. H. L. Parker, 27–37; and Garrison, *Redemptive Almsgiving in Early Christianity* 47–59.

8. Hands, *Charities and Social Aid* 81.

9. Ibid., 46; J. B. Skemp, "Service to the Needy in the Graeco-Roman World," in McCord and Parker, *Service to Christ* 17–22.

10. Quoted by Diogenes Laertius, *Lives of Eminent Philosophers* 5.21.

11. *Precepts* is a relatively late work. Ludwig Edelstein maintains that it "cannot have been written before the first century B.C. or A.D." ("The Professional Ethics of the Greek Physician," in Edelstein 322).

12. The exact antonym for *philanthropia* is *misanthropia,* "hatred of men." *Apanthropia* is "inhumanity."

13. W. H. S. Jones, ed. and trans., *Hippocrates* 1:319.

14. This is the view of Ludwig Edelstein ("The Professional Ethics of the Greek Physician" 321). It is rejected by Temkin, *Hippocrates in a World of Pagans and Christians* 31 n. 88, in favor of the view that both *philanthropia* and *philotechnia* belong to the physician. The difference in one's understanding of the passage need not affect one's view of the definition of philanthropy in *Precepts,* since *epieikea* makes it apparent that philanthropy denotes kindliness.

15. Ludwig Edelstein and Hans Diller date *On the Physician* to the fourth century B.C. at the earliest, Kudlien to the third century B.C. or later (Temkin, *Hippocrates in a World of Pagans and Christians* 23 n. 28).

16. Jones, *Hippocrates* 2:311.

17. Edelstein, "Professional Ethics" 321 f.

18. Hands, *Charities and Social Aid* 43.

19. *De Officiis* 1.14.44; Hands, *Charities and Social Aid* 47.

20. Brown, *Power and Persuasion in Late Antiquity* 95.

21. W. W. Tarn and G. T. Griffith, *Hellenistic Civilization,* 3d ed., 110.

22. Hands, *Charities and Social Aid* 85.

23. Tarn and Griffith, *Hellenistic Civilization* 109.

24. Ibid., 110.

25. *Precepts,* 4, 6, quoted by Hands, *Charities and Social Aid* 131. The word *timē* ("reputation") has a complex of meanings; e.g., losing a patient leads to loss of reputation.

26. But Natacha Massar draws attention in honorary inscriptions of the Hellenistic era to the "vertus civiques et 'morales'" (*kalokagathia*) that are assumed to accompany the professional competence of the physicians honored (*Soigner et Servir. Histoire sociale et culturelle de la médecine grecque à l'époque hellénistique* 72–3); cf. von Staden, below, n. 161.

27. H. E. Sigerist, *Civilization and Disease* 273.

28. This work is usually cited by its abbreviated Latin title, *De placitis.* The section under discussion is 9.5, in the Kühn edition 5:751 f. A critical text and English translation by Phillip De Lacy are available as vol. 5.4, 1–2 of the *Corpus Medicorum Graecorum* 565.

29. De placitis IX 5.4. Under the Roman Empire some physicians were granted exemption from certain burdensome duties including taxation: See V. Nutton, "Two Notes on Immunities: Digest 27, 1, 6, 10, and 11," *Journal of Roman Studies* 61 (1971): 52–63; reprinted in *From Democedes to Harvey* IV 52–63.

30. Menodotus, who lived in the late first and early second centuries after Christ, was leader of the Empirical school of medicine.

31. Diocles of Carystus, who lived in the fourth century B.C., was a physician known for his scientific originality.

32. Empedocles (fl. fifth century B.C.), although better known as a philosopher, was of some reputation as a physician. Galen considered him to be the founder of the Sicilian medical school.

33. Tarn and Griffith, *Hellenistic Civilization* 79–105.

34. An early and remarkable example of this feeling, perhaps from the Cynic point of view, is a poem written by Cercidas of Megalopolis (third century B.C.?), a politician who urges the upper classes to avert a social revolution by caring for the sick and giving to the needy. His attitude seems genuinely humanitarian and not merely self-serving. For a translation of the poem, see E. Barker, *From Alexander to Constantine* 58–59; see also 52.

35. See H. I. Bell, "Philanthropy in the Papyri of the Roman Period," *Hommages à Joseph Bidez et à Franz Cumont* 31–7.

36. *Attic Nights* 13.17.1.

37. Edelstein, "Professional Ethics" 336 n. 29. On the Dogmatic sect see 160 nn. 36, 37.

38. On the definition of *philanthropia*, *humanitas*, and *misericordia* as used by medical writers in the Roman imperial period, see J. Pigeaud, "Les fondements philosophiques de l'éthique médicale: Le cas de Rome," in Flashar and Jouanna, *Médecine et morale dans l'antiquité* 260–6. On *humanitas* see M. L. Clarke, *The Roman Mind* 135–45; Snell, *The Discovery of the Mind* 246–63; and Ferguson, *Moral Values in the Ancient World* 115–17.

39. Hands, *Charities and Social Aid* 87.

40. Kühn edition 1:53–63. The treatise has been translated into English by P. Brain, "Galen on the Ideal of the Physician," *South African Medical Journal* 52 (1977): 936–8.

41. R. Walzer, "New Light on Galen's Moral Philosophy (from a Recently Discovered Arabic Source)," *Classical Quarterly* 43 (1949): 82.

42. That Galen was independently well-to-do owing to his inheritance and the vast honoraria that he received from his wealthy patients undoubtedly made his *philanthropia* easier: see O. Temkin, *Galenism: Rise and Decline of a Medical Philosophy* 47, and F. Kudlien, "Medicine as a 'Liberal Art' and the Question of the Physician's Income," *JHM* 31 (1976): 453.

43. τέχνη οὕτω φιλάνθρωπος.

44. The main reason behind the refusal to be a physician to the Persian satrap is given in the pseudo-Hippocratic "letter" no. 5 in the Littré edition of Hippocrates, 9:400 ff. (for a critical text and translation see *Hippocrates: Pseudepigraphic Writings*, ed. and trans. W. D. Smith): he, as a Greek, will not treat barbarians who are enemies of the Greeks. See F. Kudlien, "Medical Ethics and Popular Ethics in Greece and Rome," *Clio Medica* 5 (1970): 94 f.

45. Temkin, *Galenism* 63.

46. J. Jouanna, "La lecture de l'éthique hippocratique chez Galien," in Flashar and Jouanna, *Médecine et morale dans l'antiquité* 211–44.

47. Temkin, *Galenism* 491.

48. Karl Deichgräber, ed., *Professio medici, Zum Vorwort des Scribonius Largus. Scribonii Largi Compositiones,* ed. Sergio Sconocchia. For an English translation of the preface, see J. S. Hamilton, "Scribonius Largus on the Medical Profession," *BHM* 60 (1986): 209–16. See also P. Mudry, "Éthique et médecine à Rome: La Préface de Scribonius Largus ou l'affirmation d'une singularité," in Flashar and Jouanna, *Médecine et morale dans l'antiquité* 297–322. Some of Largus's recipes survive in Greek.

49. Nutton, *Ancient Medicine* 174–5.

50. As Heinrich von Staden points out, however, while there are elements of Stoicism in Scribonius Largus's preface, his ideas differ in several respects from Stoic doctrine. He observes that both Scribonius Largus and Celsus avoid aligning themselves with any medical sect and that they represent a distinctive Roman point of view (in Flashar and Jouanna, *Médecine et morale dans l'antiquité* 328–9).

51. Deichgräber, *Professio medici* 24.

52. See Kudlien, "Medical Ethics and Popular Ethics" 95 f.; Edelstein, "Professional Ethics" 339, 344, and n. 45.

53. J. M. Rist, *Human Value: A Study in Ancient Philosophical Ethics;* W. den Boer, *Private Morality in Greece and Rome: Some Historical Aspects* 137–50.

54. Ferngren and Amundsen, "Virtue and Health/Medicine" 7–9.

55. H. Pétré, *Caritas: Étude sur le vocabulaire latin de la charité chrétienne* 200–21.

56. S. R. Holman, "The Entitled Poor: Human Rights Language in the Cappadocians," *Pro Ecclesia* 9 (2000): 478–9.

57. K. J. Dover, *Greek Popular Morality in the Time of Plato and Aristotle* 273–88.

58. See den Boer, *Private Morality in Greece and Rome* 128 n. 70, where den Boer refers to a study by W. Berkert. Physical deformity was regarded as a shameful thing. Thus King Croesus of Lydia did not consider his son who was deaf and dumb to be a real son (Herodotus 138.2), while the Persians were ashamed to be ruled by a governor whose ears had been cut off (Herodotus 3.73.1) (Dover, *Greek Popular Morality* 279). A defective birth (whether of a person or an animal) was regarded as a bad omen, a view that inclined the ancients against preserving it. For a comprehensive account of attitudes to physical deformity in the classical world, see R. Garland, *The Eye of the Beholder: Deformity and Disability in the Graeco-Roman World.*

59. Ferngren and Amundsen, "Virtue and Health/Medicine" 5–9. Herodotus reflects common Greek opinion when he makes Solon say to Croesus that the following blessings characterize the happy man: "He is whole of limb, a stranger to disease, free from misfortune, happy in his children, and comely to look upon" (1.32.6, trans. Rawlinson). Kudlien observes that there developed, however, a later concept of relative health in Greece in the fifth century B.C. that accepted less than perfect health as necessary to a good life (F. Kudlien, "The Old Greek Concept of 'Relative' Health," *Journal of the History of the Behavioral Sciences* 9 [1974]: 53–9).

60. On the classical attitude towards orphans see den Boer, *Private Morality in Greece and Rome* 37–61. Den Boer observes that "warmth or tenderness are [*sic*] noticeably absent from the sources of the classical period that deal with the fate of orphans" (56). In contrast, in Judaism, God was considered the father of orphans (Ps. 68:5), who were regarded as under his

protection (Mal. 3:5 and Deut. 27:19). This attitude influenced the early Christian obligation to care for orphans and foundlings.

61. K. Kapparis, *Abortion in the Ancient World* 33–52; R. Etienne, "La conscience médicale antique et la vie des enfants" 15–46.

62. One finds occasional expressions of a similar view in the classical Greek period: see, e.g., Antiphon 4, a, 2–4, quoted in Kapparis, *Abortion in the Ancient World* 169 (see 169–74 on the religious implications of abortion in the classical world).

63. Pétré, *Caritas* 175–99.

64. Rist, *Human Value* 145–52.

65. See Veyne, *Bread and Circuses* 20–1.

66. W. E. Lecky, *History of European Morals: From Augustus to Charlemagne* 1:190–1.

67. See G. Ferngren, "The *Imago Dei* and the Sanctity of Life: The Origin of an Idea," in *Euthanasia and the Newborn: Conflicts Regarding Saving Lives,* ed. R. C. McMillan, H. T. Engelhardt Jr., and S. F. Spicker, 23–45.

68. Rist, *Human Value* 129–31.

69. On the idea of the *imago Dei* in the New Testament, see *TDNT* 2:395–97, s.v. εἰκών; N. W. Porteous, "Image of God," in *The Interpreter's Dictionary of the Bible,* ed. G. A. Buttrick, 2:682–5; and W. L. Davidson, "Image of God," in *ERE* 7:160–4.

70. See *TDNT* 9:111–12, s.v. φιλανθρωπία κτλ; and Ferguson, *Moral Values in the Ancient World* 111 ff.

71. Arndt and Gingrich, *A Greek-English Lexicon* 866, s.v. φιλανθρωπία.

72. See *TDNT* 1:21–55, s.v. ἀγαπάω; Ferguson, *Moral Values in the Ancient World* 227–43; and Pétré, *Caritas* 43–61.

73. See the discussion in Hand, *Charities and Social Aid* 77–88. In the second edition of his *Byzantine Philanthropy and Social Welfare,* D. J. Constantelos argues that the gap between pagan and Christian philanthropy was not as marked as he had maintained in the first edition of his work and that Christian ideas of *philanthropia* "derived much from the mind and thought of ancient Hellenism" (ix). But while he demonstrates the existence of a philanthropic spirit in Greek philanthropy (3–13), he fails to provide evidence of a religious or moral basis for assisting the poor from which Christian conceptions of charity were derived (for a critique, see Finn 215–6). On modern approaches to classical and early Christian concepts of philanthropy, see n. 6 above.

74. On the limitations of philanthropy in pagan medical ethics, see Kudlien, "Medical Ethics and Popular Ethics" 91–7; cf. Stark 86, 88.

75. On Jewish attitudes to the poor and people in need in late antiquity and the differences between Jewish and Christian approaches, see Holman, *The Hungry Are Dying* 42–8, esp. 47–8; F. M. Loewenberg, *From Charity to Social Justice* 15–17, 181–93; and G. Hamel, *Poverty and Charity in Roman Palestine, First Three Centuries C.E.* 216–9. According to David Seccombe, there existed no organized system of Jewish public relief for the poor in Jerusalem in the Second Temple period ("Was There Organized Charity in Jerusalem before the Christians?" *Journal of Theological Studies* 29 [1978]: 140–3).

76. *ANF* 8:285. Cf. the very similar passage (to which this is clearly related) in the *Recognitions* of Clement 5.23 (*ANF* 8:148–9).

77. On the lack in classical ethics of a concept of neighborly love that extended to all

people, see den Boer, *Private Morality in Greece and Rome* 62–72, and Downey, "Who Is My Neighbor?" 2–15.

78. J. Agrimi and C. Crisciani, "Charity and Aid in Medieval Christian Civilization," in Grmek, *Western Medical Thought* 170–2.

79. Downey, "Who Is My Neighbor?" 14; D. F. Winslow, "Gregory of Nazianzus and Love for the Poor," *Anglican Theological Review* 47 [1965]: 14. See, however, Temkin, *Hippocrates in a World of Pagans and Christians* 254.

80. H. G. Liddell and R. Scott, *A Greek-English Lexicon* (9th ed.) 657, s.v. ἐπισκοπέω. The same verb is also used in James 1:27.

81. Veyne calls Christian mutual assistance and almsgiving "a sectarian morality" to differentiate it from the pagan practice of philanthropy that arose from euergetism (*Bread and Circuses* 23).

82. For Latin synonyms of *philanthropia* used by the Latin fathers, see Pétré, *Caritas* 62–100.

83. G. Downey, "Philanthropia in Religion and Statecraft in the Fourth Century after Christ," *Historia* 4 (1955): 199–208, esp. 204 ff. On the change in the Christian definition of philanthropy in late antiquity, see Finn 191–7. "Almsgiving was incorporated into a moral theology of the virtues, in particular as an expression of *philanthropia, liberalitas, humanitas,* and *iustitia.* These virtues brought to almsgiving a new honor and status for practitioners" (Finn 264); cf. Garrison, *Redemptive Almsgiving in Early Christianity* 76–108 (though he is not convincing in his thesis that early Christianity, beginning with the New Testament, encourages a belief in redemptive almsgiving). On different classical and Christian meanings of *philanthropia,* see Finn 214–20, and E. Patlagean, *Pauvreté économique et pauvreté sociale à Byzance 4e–7e siècles* 181–96.

84. On the influence of *humanitas* in the practice of Roman medicine as it influenced attitudes to such matters as chronic disease, the value of life, suffering, and vivisection, see Pigeaud, "Les Fondements philosphiques de l'éthique médicale: Le Cas de Rome" 266–90. For an extensive bibliography of recent works on exposure and infanticide, see E. A. Castelli, "Gender, Theory, and *The Rise of Christianity*: A Response to Rodney Stark," *Journal of Early Christian Studies* 6 (1998): 236 n. 16.

85. The two most notable instances are the Hippocratic Oath, which forbids giving a pessary to a woman to procure an abortion (for the interpretation of this highly controverted stipulation, see Kapparis, *Abortion in the Ancient World* 66–76), and an inscription from Philadelphia in Lydia, which prohibits inter alia the knowledge, use, or dissemination of abortifacient drugs and techniques (for a translation with commentary and bibliography, see ibid., 214–8). Dating from the second or first century B.C., the inscription regulates entry to a private sanctuary and sets forth a moral standard that Kapparis thinks anticipates Christian views.

86. See Kapparis, *Abortion in the Ancient World* 167–94; on the rescript (an imperial ruling in response to a petition or judicial enquiry), which probably reflects Roman concerns regarding a declining population, see ibid., 182–5.

87. See especially Justin, *Apologia* 1.27, 29; Tertullian, *Ad nationes* 1.16; idem, *Apologia* 9.17–18; Clement of Alexandria, *Paedagogus* 3.3.21.5; idem, *Stromateis* 2.18.92–93, 5.14; Minucius Felix, *Octavius* 30.2, 31.4; *Epistle to Diognetus* 5.6; Athenagoras, *Supplicatio* 35.6; and

Origen, *Contra Celsum* 8.55. I owe these citations to Castelli, "Gender, Theory, and *The Rise of Christianity*" 237 n. 20. For a discussion of these passages and other early Christian writers that deal with abortion, see A. Lindemann, "'Do Not Let a Woman Destroy the Unborn Baby in Her Belly': Abortion in Ancient Judaism and Christianity," *Studia Theologica* 49 (1995): 253–71; J. T. Noonan Jr., "An Almost Absolute Value in History," in *The Morality of Abortion: Legal and Historical Perspectives,* ed. J. T. Noonan Jr., 7–18; and J. Connery, *Abortion: The Development of the Roman Catholic Perspective* 33–64.

88. For a bibliography of recent discussions of abortion in the classical world, see Castelli, "Gender, Theory, and The Rise of Christianity" 239 n. 24. John Riddle has argued that contraception was much more widely practiced in the classical world than had previously been thought (*Eve's Herbs: A History of Contraception and Abortion in the West*). For a critique of this view, as a highly speculative one that is built on slender data, see G. Ferngren, Review of John M. Riddle, *Eve's Herbs: A History of Contraception and Abortion in the West, New England Journal of Medicine* 337 (November 6, 1997): 1398.

89. "We put down mad dogs; we kill the wild, untamed ox; we use the knife on sick sheep to stop their infecting the flock; we destroy abnormal offspring at birth; children, too, if they are born weak or deformed, we drown. Yet this is not the work of anger, but of reason—to separate the sound from the worthless" (Seneca, *De ira* 1.15.2, trans. and ed. by J. M. Cooper and J. F. Procopé, in Seneca, *Moral and Political Essays* 32). On physicians' attitudes to abortion in the classical world, see Kapparis, *Abortion in the Ancient World* 53–89.

90. *Apologeticum ad nationes* 1.15. Quoted in Noonan, "An Almost Absolute Value in History" 12. Kapparis argues that the lack of identification of abortion with homicide in pagan values was the result of a failure to agree on whether the unborn were human *(Abortion in the Ancient World* 174).

91. Amundsen, "Medicine and the Birth of Defective Children," in Amundsen, 50–69.

92. For a brief survey of physical disability or deformity as reason for exposure, see Garland, *The Eye of the Beholder* 11–18, and Kapparis, *Abortion in the Ancient World* (who surveys the recent literature) 54–62. The exposure of children was formally condemned in law in A.D. 374 (*Corpus Juris Civilis, II, Codex Justinianus* 8, 51, 2 and 9, 16, 7). Roman law regarding abortion is discussed by Connery, *Abortion* 22–32.

93. See, e.g., Minucius Felix, *Octavius* 30.2 and 31.4, and Lactantius 6.20.

94. On the ancient physician's willingness to assist in suicide, see Amundsen, "The Physician's Obligation to Prolong Life" 37–40, and Gourevitch, *Le Triangle hippocratique* 160–216.

95. *City of God* 1. 22–27.

96. D. W. Amundsen, "Suicide and Early Christian Values," in *Suicide and Euthanasia: Historical and Contemporary Themes,* ed. B. A. Brody, 77–153; reprinted in Amundsen 70–126.

97. For what follows I am indebted to G. G. Stroumsa, "*Caro salutis cardo*: Shaping the Person in Early Christian Thought," *History of Religions* 30 (1990): 25–50; cf. Dörnemann, *Krankheit und Heilung* 299–307.

98. Stroumsa, "*Caro salutis cardo*" 37.

99. Ibid., 40, 41.

100. C. W. Bynum, *The Resurrection of the Body in Western Christianity, 200–1336* 11 n. 17.

101. Ibid., 11; Stroumsa, "*Caro salutis cardo*" 39–40.

102. Stroumsa, "*Caro salutis cardo*" 40.

103. Ibid., 38.

104. Ibid., 38–9.

105. Ibid., 30.

106. Ibid., 35.

107. Ibid., 30.

108. So Irenaeus, *Adversus Haereses* 5.14.1. Stroumsa, "*Caro salutis cardo*" 42.

109. So Athenagoras, *De resurrectione* 15.2, 15.6.

110. Stroumsa, "*Caro salutis cardo*" 37.

111. Ibid., 47.

112. Ibid., 47, 38.

113. Ibid., 49.

114. Ibid., 49–50. For the implications of this novel understanding for philanthropy, in which a love of the poor came to replace an attachment to the community, see ch. 6, pp. 121–3.

115. Stroumsa, "*Caro salutis cardo*" 35–6. Cf. Eph. 5:30 for a similar analogy.

116. Stroumsa, "*Caro salutis cardo*" 44.

117. Holman, *The Hungry Are Dying* 149–51. Holman describes the belief of Basil, Gregory of Nazianzus, and Gregory of Nyssa that the poor deserve compassion both on the basis of a common kinship as humans (148–9) and as those who share in the divine image, particularly with the incarnate nature of Christ, whom they signify. No part of the body is dishonorable or evil: even the reproductive organs have a divine purpose in preserving humanity (Gregory of Nyssa, *Catechetical Oration* 28 as quoted in ibid., 164–5).

118. The definition of the poor as a separate category was peculiar to Jewish and Christian charity but lacking in pagan philanthropy (see Veyne, *Bread and Circuses* 30; Winslow, "Gregory of Nazianzus and Love for the Poor" 348–59; Brown, *Power and Persuasion in Late Antiquity* 91–3; and idem, *Poverty and Leadership* 1–44, 69–71).

119. Holman, *The Hungry Are Dying* 27–29, 181.

120. Ibid., 152.

121. *Paup.* 2.

122. Holman, *The Hungry Are Dying* 158–60. On the Cappadocians' understanding of the leper's body, see 160–6.

123. Ibid., 167.

124. See above, ch. 5, pp. 87, 89–90.

125. Brown, *Poverty and Leadership* 92–93. For a social and rhetorical analysis of the Cappadocians' sermons, see Holman, "The Entitled Poor" 476–89.

126. Brown, *Poverty and Leadership* 94.

127. Ibid., 95–6.

128. Ibid., 104. The motif of human solidarity plays an important role in Gregory of Nazianzus's sermon περὶ πτωχοτροφίας ("On the Feeding of the Poor"), in which Gregory asserts that we strengthen the image of God in man by ministering to the bodily needs of the poor while we are in danger of destroying that image when we turn our back on them; the beggar may better preserve the image of God in his diseased body than do people who lack

charity, where disease infects the soul (chs. 14 and 18) (Winslow, "Gregory of Nazianzus and Love for the Poor" 352). On the use of sickness and healing as metaphors by Gregory Nazianzus and Gregory of Nyssa, see Dörnemann, *Krankheit und Heilung* 219–73.

129. Brown, *Poverty and Leadership* 108, 111.

130. Ibid., 111; cf. Winslow, "Gregory of Nazianzus and Love for the Poor" 353–4. Gregory Nazianzus speaks of acts of mercy as imitating God (περὶ πτωχοτροφίας chs. 24–27). Winslow sees here the assertion of a "divine-human solidarity" that both echoes classical ideas of emulating the gods by showing acts of kindness and goes beyond them by grounding compassion in the Incarnation of Christ, so that one becomes like him (353–5).

131. See ch. 2, pp. 29–31.

132. Pease, "Medical Allusions" 74–5.

133. Ibid., 75.

134. V. Nutton, "God, Galen and the Depaganization of Ancient Medicine," in *Religion and Medicine in the Middle Ages,* ed. P. Biller and J. Ziegler, 22.

135. R. Grant, "Paul, Galen and Origen," *Journal of Theological Studies* 34 (1983): 535.

136. Pease, "Medical Allusions" 81–2 (citing Lübeck).

137. Eusebius, *Eccles. Hist.* 5.28.13–14; see R. Walzer, *Galen on Jews and Christians* 77–86; R. L. Wilken, *The Christians As the Romans Saw Them* 79–83; S. Benko, *Pagan Rome and the Early Christians* 142–5. Eusebius describes Theodotus as the first person to hold the view that Jesus was only a man. These Christians were later excommunicated for heresy.

138. On Christian attacks on Galen, see Temkin, *Hippocrates in a World of Pagans and Christians* 203–6. On Galen's influence on Christians, see Bernard, "Athenagoras" 15–16.

139. See W. A. Fitzgerald, "Medical Men: Canonized Saints," *BHM* 22 (1948): 635–41. J. A. Pattengale ("Benevolent Physicians in Late Antiquity: The Cult of the Anargyroi" 114–36) lists a number of Christian physicians from the first five centuries of Christianity (see esp. 119). Many are taken from physician martyrologies and (as also in the case of the physicians listed by Fitzgerald) some are semi-legendary.

140. He is called "the beloved physician" (ὁ ἰατρὸς ὁ ἀγαπητὸς) in Colossians 4:14. The tradition that Luke was a gentile physician from Antioch is maintained by several later sources including the anti-Marcionite Prologue and the Muratorian Canon, both of which date from the late second century. W. K. Hobart's thesis that the Third Gospel is suffused with first-century medical terminology (*The Medical Language of St. Luke* [1882]) was challenged by H. J. Cadbury (*The Style and Literary Method of Luke* [1919–20] 39–72). For a summary of the debate and subsequent literature, see C. J. Hemer, *The Book of Acts in the Setting of Hellenistic History,* ed. Conrad H. Gempf 308–12. While the linguistic evidence for the use of medical terminology is inconclusive, there is no reason to doubt the Colossian identification of Luke as a physician.

141. Eusebius, *Eccles. Hist.* 5.1.49. There is no reason to assume that Alexander "worked miracles of healing" (so Lane Fox, *Pagans and Christians* 329) as a physician except on the assumption that his being credited with "the charisma of the Apostles" implies that he had the spiritual gift of healing.

142. Eusebius, *Eccles. Hist.* 8.13.4. He calls him the "best of physicians." He also speaks of one Theodotos, bishop of Antioch in the third century, who was skilled in the art of healing both bodies and souls, which suggests that he was trained as a physician (7.32).

143. C. Schulze, *Medizin und Christentum in Spatäntike und frühem Mittelalter: Christliche Ärzte und ihr Wirken.* Schulze incorporates the work of earlier scholars and discusses separately the source of each physician identified, some of whom are anonymous.

144. The number represents Christian physicians of unquestionable status; it is increased to 194 if one includes names whose status is not certain. A convenient list of all 194 can be found on pp. 235–39.

145. Schulze, *Medizin und Christentum* 144–50.

146. "Christentum und Medizin gingen in der alltäglichen antike Praxis, bei der Berufsausübung bekannter und unbekannter Männer und Frauen, offenbar problemlos Hand in Hand....daß die ärztliche Tätigkeit von Christen nicht unterdurchschnittlich oft ausgeübt wurde, sonder daß sie, ganz im Gegenteil, sogar zu den besonders beliebten Berufen gezählt haben dürfte" (ibid., 154).

147. See above, n. 2.

148. Peter Brown points out that Christianity by the fourth century was prominent especially in urban areas in the eastern provinces, with many socially visible leaders (*Poverty and Leadership* 17).

149. In another study Christian Schulze has assembled evidence of Christian women who were physicians. He identifies eight of seventy-nine physicians as women and finds that they enjoyed a comparatively higher status than did pagan female physicians (C. Schulze, "Christliche Ärztinnen in der Antike," in Schulze and Ihm, *Ärztekunst und Gottvertrauen* 91–115).

150. Temkin, *Hippocrates in a World of Pagans and Christians* 186–7; on the extent of physicians' relationship to Asclepius, see ibid., 181–96.

151. Most of the examples that Nutton cites that depict physicians as ambiguous figures are taken from late antiquity, the Byzantine period, or the Middle Ages ("God, Galen and Origen" 19–20). It is interesting that in the list of forbidden occupations given by Hippolytus (*Apostolic Tradition* 16) physicians are not mentioned in any of the several versions that we have (G. J. Cuming, *Hippolytus: A Text for Students with Introduction, Translation, Commentary and Notes* 15–16).

152. Galen recognized that only exceptional physicians possessed the ability to become philosophers (*That the Best Physician Is Also a Philosopher,* passim). He acknowledged that good men could practice medicine by using empirical methods.

153. *Contra Celsum* 3.74.

154. Sermones 175.8 ff.

155. On the influence of new emphases in Christian ethics on the larger Roman society, see Veyne, *Bread and Circuses* 24–6.

156. Temkin, *Hippocrates in a World of Pagans and Christians* 252–4. On the ambiguities that faced the Christian physician in the overlapping of incompatible Christian and classical medical ethics, see Kudlien, "Medical Ethics and Popular Ethics" 97.

157. Epistles 159.

158. Constantelos, *Byzantine Philanthropy and Social Welfare* 182.

159. Sermones 9.10.

160. *In Ps.* 125.14.

161. Heinrich von Staden adduces evidence of the continuity that existed between the Hellenistic and Roman imperial periods in emphasizing both professional competence and

moral probity as characteristics of the good physician ("Character and Competence: Personal and Professional Conduct in Greek Medicine," in Flashar and Jouanna, *Médecine et morale dans l'antiquité* 157–72).

162. B. A. Lustig, "Compassion," in *The Encyclopedia of Bioethics,* ed. W. T. Reich, 2d ed., 1:440.

163. Pétré, *Caritas* 222–39.

164. Finn 189.

165. Finn argues against Pétré, *Caritas* 251, that the late Christian understanding of *misericordia* sought to "integrate emotion and action" by focusing on motive and intention as well as the act of almsgiving itself (Finn 189–90).

166. Callinicus, *Life of Hypatios* 22.2; see the remarks of Constantelos, *Byzantine Philanthropy and Social Welfare* 95–6, and of Finn 189.

167. Epistle 189.

168. E.g., *The Rule of St. Benedict* 31; Cassiodorus, *Institutes of Divine and Human Readings* 1.31.

169. W. H. S. Jones, *The Doctor's Oath: An Essay in the History of Medicine* 50.

170. On the oath, see L. Edelstein, *The Hippocratic Oath: Text, Translation, and Interpretation,* supplements to the *Bulletin of the History of Medicine* 1, reprinted in Edelstein 3–63; C. Lichtenthaeler, *Der Eid des Hippokrates*; Nutton, "Beyond the Hippocratic Oath" 19–37; K. Deichgräber, *Der hippokratische Eid.*; H. von Staden, "In a Pure and Holy Way: Personal and Professional Conduct in the Hippocratic Oath," *JHM* 51 (1996): 404–37; von Staden, "Character and Competence" 172–95; T. Rütten, "Medizenethische Themen in den deontologischen Schriften des *Corpus Hippocraticum,*" in Flashar and Jouanna, *Médecine et morale dans l'antiquité* 68–98; C. Schubert, *Der hippokratische Eid. Medizin und Ethik von der Antike bis heute*; and T. Rütten, *Geschichten vom Hippokratischen Eid.*

171. Vivian Nutton dates the original oath to the period 410–350 B.C. ("Beyond the Hippocratic Oath" 28 n. 2).

172. Edelstein, *Hippocratic Oath* 17–20. The suggestion has not met with wide acceptance. On the debate regarding Pythagorean influence, see Kapparis, *Abortion in the Ancient World* 66–76 passim.

173. It may be for this reason that Caesarius, brother of Gregory of Nazianzus, a physician who studied medicine in Alexandria, refused to swear the oath (Greg. Naz., *Oratio* VII, art. 10; see Temkin, *Hippocrates in a World of Pagans and Christians* 182 and n. 8). Vivian Nutton thinks that the oath became a sign of anti-Christian behavior in fourth-century Alexandria.

174. Edelstein, "Professional Ethics," 331 and n. 22. The extent to which the oath was sworn by physicians is disputed: see Nutton, "Beyond the Hippocratic Oath" 20 and 23–4.

175. Jerome, writing in the late fourth century, has been thought to cite the oath with approval (*Epistulae* 52.15). He is likely, however, to be attributing to the oath precepts drawn from other Hippocratic deontological works (L. C. MacKinney, "Medical Ethics and Etiquette in the Early Middle Ages: The Persistence of Hippocratic Ideals," *BHM* 26 [1952]: 3–4; C. R. Galvão-Sobrinho, "Hippocratic Ideals, Medical Ethics, and the Practice of Medicine in the Early Middle Ages: The Legacy of the Hippocratic Oath," *JHM* 51 [1996]: 440–2). Even if Jerome did not make use of the oath, his interest in Hippocratic treatises testifies to his sympathy with some elements of Greek medical ethics.

176. See Jones, *The Doctor's Oath* 17–27, for texts with critical apparatus and translations of several versions, with discussion on 51–55. Jones thinks that the Christian oath antedates the time of Galen (55), but this view is highly speculative. Lichtenthaeler and Deichgräber date the original Oath to the fifth century B.C., which is plausible (Kapparis, *Abortion in the Ancient World* 70).

177. Galvão-Sobrinho, "Hippocratic Ideals" 446–53; Temkin, *Hippocrates in a World of Pagans and Christians* 247–8.

178. This document is preserved in *Variae* 6.19.

179. *Institutiones* 1.31.

180. MacKinney, "Medical Ethics and Etiquette" 1–31.

181. In *De clem.* 2.6–7 Seneca draws a distinction between clemency, which he considers one of the highest virtues, and pity, which he considers a vice (see, on Seneca, B. F. Harris, "The Idea of Mercy and Its Graeco-Roman Context," in *God Who Is Rich in Mercy: Essays Presented to D. B. Knox,* ed. P. T. O'Brien and D. G. Peterson, 98–99. Holman remarks on the "absence of 'pity' language" in classical texts (*The Hungry Are Dying* 11). "As Christian discussion of philanthropy changed the focus of 'good deeds' from the city to the bodies of the poor," she writes, "it redefined the criteria for receiving welfare. Physical need now mattered for its own sake and not solely in terms of civic order" (18). See also E. A. Judge, "The Quest for Mercy in Late Antiquity," in O'Brien and Peterson, *God Who Is Rich in Mercy* 107–21.

182. See, e.g., Castelli for an extensive modern bibliography that deals with the influence of Christian asceticism on "reinscribing" gender hierarchies ("Gender, Theory, and *The Rise of Christianity*" 251 and n. 54; cf. Lecky, *History of European Morals* 2:361–3). The classical rather than the Christian view of the poor can also be found in texts from late antiquity, indicating that some Christians continued to feel more comfortable with the Greek view, which did not consider poverty to confer a specific social identity. Holman cites Synesius (c. 370–c. 414), a pupil of Hypatia and bishop of Pentapolis in Cyrene, as an example (*The Hungry Are Dying* 177–9).

183. Holman, *The Hungry Are Dying* 180.

Chapter 6 • Health Care in the Early Church

1. Keith Hopkins argues that the greatest period of Christian growth was in the third century, when the number of Christians grew from about 200,000 to more than 6 million ("Christian Number and Its Implications" 192–3, 198).

2. See, e.g., Hippolytus, *Apos. trad.*, Canon 20. C. E. B. Cranfield, "Diakonia in the New Testament," in McCord and Parker, *Service to Christ* 37–48; R. Grant, "The Organization of Alms," in *Early Christianity and Society: Seven Studies* 124–45; Finn 41–56.

3. *Apost. const.* 3.19 (quoted in Stark 87); cf. Polycarp, *Ep. ad Phil.* 6.1. See J. Colson, *La Fonction diaconale aux origines de l'Église* 9–120, for a thorough discussion of the New Testament and subapostolic periods; J. G. Davies, "Deacons, Deaconesses and the Minor Orders in the Patristic Period," *Journal of Ecclesiastical History* 14 (1963): 1–6; and G. W. H. Lampe, "Diakonia in the Early Church," in McCord and Parker, *Service to Christ* 49–64.

4. Justin Martyr, *Apology* 1.67.

5. Tertullian, *Ad uxor.* 2.4. On deaconesses, see Colson, *La Fonction diaconale* 121–39, and

Schulze, "Christliche Ärztinnen in der Antike" 91–115. Davies suggests that deaconesses came into existence as a separate order in the first half of the third century ("Deacons" 2). Crislip argues that women were assigned a traditional role in Greek society in caring for the sick (see Crislip 44 and 165–6 n. 25), but this is denied by Helen King ("Using the Past: Nursing and the Medical Profession in Ancient Greece," in *Anthropology and Nursing*, ed. P. Holden and J. Littlewood, 10–14). For a comparison of pagan and Christian caregivers and the extent to which gender roles played a part, see C. Schweikardt and C. Schulze, "Facetten antiker Krankenpflege und ihrer Rezeption," in Schulze and Ihm, *Ärztekunst und Gottvertrauen* 117–38. The authors argue that pagan caregivers were men, while Christian caregivers were women (134).

6. *Eccles. Hist.* 6.43.

7. On the basis of these figures Robert Grant has conservatively estimated that there were between 15,000 and 20,000 Christians in Rome in 251 ("The Christian Population of the Roman Empire," in *Early Christianity and Society* 7). Other estimates place the Christian population of Rome at twice that number. On the development of minor orders of clergy, see Davies, "Deacons" 6–15.

8. Harnack, *Mission and Expansion* 1:195 n. 1.

9. Tertullian, *De praescr. haeret.* 30; *Adv. Marc.* 4.4.

10. Epistle 62.

11. *Homil. in Matt.* 66/67.3.

12. G. C. Kohn, "Plague of Cyprian," in *Encyclopedia of Plague and Pestilence,* ed. G. C. Kohn, 250–1.

13. For a description of the symptoms of the plague in Carthage, see Cyprian, *De mortalitate* 14.

14. *New History* 1.26 and 37; cf. also 36 and 46.

15. *Hist. Aug., Vita Gall.* 5.6. For a discussion of the rate of mortality, see Stark 76–7. Stark cites several estimates but follows McNeill in assuming a mortality of from 25 to 33 percent.

16. On sanitation in the cities of the Roman Empire, see A. Scobie, "Slums, Sanitation, and Mortality in the Roman World," *Klio* 68 (1986): 407–22.

17. R. Jackson, *Doctors and Diseases in the Roman Empire* 52–3.

18. Scobie, "Slums, Sanitation, and Mortality" 421 and n. 109 for citations of sources; Patlagean, *Pauvreté économique et pauvreté sociale* 101–12.

19. Vivian Nutton observes that medicine in classical times was poorly equipped to deal with plague ("Healers in the Medical Market Place" 33).

20. On the plague of Justinian, see L. K. Little, ed., *Plague and the End of Antiquity: The Pandemic of 541–750.*

21. G. F. Gilliam observed that "descriptions of pestilence in any period are likely to be highly coloured and extravagant" (quoted by P. Allen, "The 'Justinianic' Plague," *Byzantion* 49 [1979]: 10).

22. *History* 2.51–3.

23. Edelstein and Edelstein, *Asclepius* 174–5.

24. Scobie, "Slums, Sanitation, and Mortality" 431.

25. On the morbidity and mortality of famine and other catastrophes in late antiquity, see Patlagean, *Pauvreté économique et pauvreté sociale* 74–92, and Jones, *The Later Roman Empire*

2:810–1, 853–5. For the famine at Edessa (below and n. 26), see P. Garnsey, *Famine and Food Supply in the Graeco-Roman World: Responses to Risk and Crisis* 20–37, who uses it as a case study and compares it with similar catastrophes in the classical world. Garnsey argues that while food crises were common, famines were rare (6). On the limitations of intervention by cities and the Roman imperial government in crisis situations, see Garnsey, *Famine and Food Supply* 257–68 and passim. On the social dynamics of famine, see S. R. Holman, "The Hungry Body: Famine, Poverty, and Identity in Basil's *Hom.* 8," *Journal of Early Christian Studies* 7, no. 3 (1999): 353–61.

26. Eusebius, *Eccles. Hist.* 9.8.

27. Stark 206. The italics are Stark's.

28. Cf. ibid., 86–7.

29. Sozomen, *Eccles. Hist.* 3.16. On the career of Ephraem, see S. H. Griffith, "Ephraem, the Deacon of Edessa, and the Church of the Empire," in *Diakonia: Studies in Honor of Robert T. Meyer*, ed. T. Halton and J. Williman 22–52.

30. Joshua Stylites, *Chron.* 26.28.41–3. For a convenient translation of ps. Joshua the Stylite's description of the plague, see Garnsey, *Famine and Food Supply* 3–6. Edessa enjoyed a reputation both for distinguished Christian physicians and for medical philanthropy. Rabula, bishop of the city from 412 to 435, established permanent hospitals (*nosokomeia*) each for men and women, doubtless drawing on earlier experiences of providing temporary shelters (J. B. Segal, *Edessa "The Blessed City"* 71 and 148).

31. Writing in 260, Dioysius says that the population of Alexandria between the ages of 40 and 70 before the plague was larger than the number of inhabitants between 14 and 80 after the plague who were eligible to receive a distribution of public grain (Eusebius, *Eccles. Hist.* 7.21.9).

32. Quoted by Eusebius, *Eccles. Hist.* 7.22.7.

33. Pontius, *Vita Cypriani* 9; Cyprian, *Ad Demetrianum* 10–11.

34. *Ad Demetrianum* 2.

35. *Apology* 40. By abandoning the worship of Rome's traditional gods in favor of the novel worship of a strange god, the Christians (so they believed) had forfeited the *pax deorum*, the divine protection on which Rome depended for her security, and so deserved retribution. Jacob Burckhardt observed that there existed throughout antiquity "a minimum of belief in the gods" that required prudence whenever the safety of the state was involved (*The Greeks and Greek Civilization*, trans. S. Stern and ed. O. Murray, 272–3).

36. Pontius, *Vita Cypriani* 9–11.

37. W. Smith and H. Wace, *A Dictionary of Christian Biography* 1:747, s.v. "Cyprianus (1) Thascius Caecilius."

38. *Vita Greg. Thaumat.* 12.

39. Arthur Boak calculated that the Egyptian town of Karanis suffered the loss of more than one-third of its population (Stark 77).

40. Quoted by Eusebius, *Eccles. Hist.* 7.22.

41. Tertullian, *Apol.* 39.6; Aristides, *Apol.* 15. *DCA* 1:459 and 684, s.v. "Copiatae," "Fossarii." See É. Rebillard, "Les Formes de l'assistance funéraire dans l'empire romain et leur évolution dans l'antiquité tardive," *Antiquité tardive* 7 (1999): 275–8.

42. *Eccles. Hist.* 9.8.

43. Optatus Milevitanus, *On the Schism of the Donatists*, app. 1.

44. Brown, *Power and Persuasion in Late Antiquity* 103 with sources cited in nn. 172 and 173.

45. Ibid., 102 with sources cited in n. 167. See É. Rebillard, "Église et sépulture dans l'Antiquité tardive (Occident latin, 3e-6e siècles)," *Annales HSS* 54:5 (1999): 1027–46.

46. *Epist. ad Arsac.* 49.

47. Gibbon, *The Decline and Fall of the Roman Empire* 5:115 n. 24.

48. On the *parabalani,* see A. Philipsborn, "La Compagnie d'Ambulanciers 'Parabalini' d'Alexandrie," *Byzantion* 20 (1950): 185–90; W. Schubart, "Parabalani," *Journal of Egyptian Archaeology* 40 (1954): 97–101; F. Cabrol and H. Leclercq, eds., *Dictionnaire d'archeologie chrétienne et de liturgie,* s.v. "Parabalani," 13/2, cols. 1574–8, by H. Leclercq; *DCA* 2:1551–2, s.v. "Parabolani"; and Brown, *Power and Persuasion in Late Antiquity* 102.

49. On the violence that was a feature of the large cities of the Roman Empire (especially Rome, Antioch, and Alexandria) and in which groups like the *parabalani* and *fossores* became involved, sometimes as agents of their bishops, see Brown, *Power and Persuasion in Late Antiquity* 87–9, 102–3, 107–8.

50. Socrates, *Eccles. Hist.* 7.15. Mobs composed of Christian monks came to be used by some bishops in large Roman cities to undertake violent attacks on pagan temples and other buildings in the late fourth century (see Brown, *Power and Persuasion in Late Antiquity* 103–17).

51. *DCA* 2:1551–2, s.v. "Parabolani."

52. *Eccles. Hist.* 9.8.

53. Stark 74–5. The italics are Stark's.

54. See his discussion of differential mortality (ibid., 88–91); quotation at 89. The italics are Stark's.

55. Ibid., 91–4.

56. Suggested by Philipsborn, "La Compagnie d'Ambulanciers" 185–90. While this is plausible, it cannot be proved.

57. Brown, *Poverty and Leadership* 1–44, especially 3.

58. Brown argues that the new Christian emphasis on the love of the poor introduced into classical society a Near Eastern definition, in which the poor were defined judicially rather than economically as a class of "the weak" who sought justice from a superior, in this case the bishop, who emerged as the protector of the weak and distressed ("the poor"). It is the picture of the poor in the Hebrew scriptures (ibid., 69–71).

59. In Greek, different words were employed to indicate two kinds of poverty: *penēs,* which denoted a member of the working poor, and *ptōchos,* which denoted one who was impoverished and depended on others for support (E. Patlagean, "The Poor," in *The Byzantines,* ed. G. Cavallo, trans. T. Dunlap, T. L. Fagan, and C. Lambert, 15–16; cf. Arndt and Gingrich, *A Greek-English Lexicon,* s.v. πένης and πτωχός). The latter (πτωχός) came to figure in the language of Christian philanthropy.

60. Patlagean, *Pauvreté économique et pauvreté sociale* 185–8, 231–5 and 423–32; idem, "The Poor" 18–19.

61. *Brown, Poverty and Leadership* 8, 91.

62. It is worth noting that the care of the poor extended only to the *urban* poor; those

who lived outside the cities did not enjoy proximity to Christian institutional charity (ibid., 50–1).

63. Finn 182–8.

64. Brown, *Poverty and Leadership* 31; on the institutionalization of Christian charity, see P. Garnsey and C. Humfress, *The Evolution of the Late Antique World* 123–31.

65. A. H. M. Jones argues that Christianity chiefly appealed to the poor and lower classes in the cities of the fourth century ("The Social Background of the Struggle between Paganism and Christianity," in Momigliano, *The Conflict between Paganism and Christianity in the Fourth Century* 17–37).

66. Finn 32.

67. Ibid., 108–15.

68. *Homiliae in Matthaeum* 66.3 (*Patrologia Graeca* 58:630); Brown, *Power and Persuasion in Late Antiquity* 94.

69. For the sources, see Brown, *Power and Persuasion in Late Antiquity* 98 n. 146.

70. On the term "Christianization of euergetism," see Brown, *Poverty and Leadership* 77; on Christian euergetism as a continuation of pagan euergetism, see Garnsey and Humfress, *Evolution of the Antique World* 115–23. Brown argues (41–42) that a quid pro quo underlay Basil's famine relief that preceded the founding of the Basileias: in return for a commitment to assuming famine relief, Basil was granted certain privileges by Constantius II, which would have included remission of taxes. He emerged as "the new-style euergetes of a Christian city in its hour of need" and helped the emperor maintain the threatened social fabric (42). The evidence is circumstantial, and Brown's reconstruction depends on dating Basil's famine relief at 370.

71. Three well-known "poverty sermons" were composed by Gregory of Nazianzus and Gregory of Nyssa. They are the former's Oration 14, *peri philoptōchias* (On the Love of the Poor), also known as *peri ptōchotrophias* (On the Feeding of the Poor), composed between A.D. 365 and 372; and the latter's two sermons *De pauperibus amandis* (On the Love of the Poor), composed between A.D. 372 and 382. The sermons are of particular interest because they provide the ideological framework for the establishment of Basil's hospital (see Holman, "Healing the Social Leper" 283–309).

72. Susan Holman observes that the Cappadocians incorporated the values of classical culture into their rhetoric of poor relief by appropriating already-existing concepts, such as patronage, kinship, and the exchange of gifts. In so doing they created a new civic identity for the poor ("The Entitled Poor" 481).

73. Patlagean, "The Poor" 20.

74. Ibid., 89, 1, and 45. John Chrysostom saw himself as the ambassador of the poor (*On Alms* 1).

75. Brown, *Power and Persuasion in Late Antiquity* 99.

76. Cf. ibid., 8 and 110–1.

77. On the origin of the hospital, see G. E. Gask and J. Todd, "The Origin of Hospitals," in *Science, Medicine, and History: Essays on the Evolution of Scientific Thought and Medical Practice Written in Honour of Charles Singer*, ed. E. A. Underwood, 1:122–30; T. S. Miller, *The Birth of the Hospital in the Byzantine Empire*, 2d ed.; G. B. Risse, *Mending Bodies, Saving Souls: A History of Hospitals* 69–116; N. Allan, "Hospice to Hospital in the Near East: An Instance of

Continuity and Change in Late Antiquity," *BHM* 64 (1990): 446–62; G. Harig, "Zum Problem 'Krankenhaus' in der Antike," *Klio* 53 (1971): 179–95; and Crislip 100–42.

78. Most recent historians of the hospital are agreed on this point: see Gask and Todd, "The Origins of Hospitals" 122–5; Miller, *The Birth of the Hospital* 30–49; Risse, *Mending Bodies, Saving Souls* 38–59; Crislip 120–8; and P. Horden, "The Christian Hospital in Late Antiquity: Break or Bridge?" in *Gesundheit—Krankheit. Kulturtransfer medizinischen Wissens von der Späntantike bis in die Frühe Neuzeit,* ed. F. Steger and K. P. Jankrift, 88, 92. Loewenberg has found no examples in either the Hebrew scriptures or Talmudic literature of Jewish institutional care of the sick (Loewenberg, *From Charity to Social Justice* 146–7); cf. Horden, "Christian Hospital in Late Antiquity" 90–2.

79. *Valetudenaria* for both slaves and soldiers declined in the third century and eventually disappeared altogether as *coloni* replaced slaves and the Roman army was reorganized (Horden, "Christian Hospital in Late Antiquity" 89–90).

80. V. Nutton, Essay Review of *The Birth of the Hospital in the Byzantine Empire* by Timothy S. Miller, *Medical History* 30 (1986): 219.

81. J. N. D. Kelly, *Golden Mouth: The Story of John Chrysostom, Ascetic, Preacher, Bishop* 120.

82. Miller, *The Birth of the Hospital* 23–9. The fact that the names of many of the Christian institutions are neologisms in both Greek and Latin indicates their novelty (Veyne, *Bread and Circuses* 33 and 63 n. 45).

83. Nutton, Essay Review 219.

84. Crislip 101–2.

85. Basil, Epistles 94, 150, 176; Gregory of Nazianzus, Oration 43.61–63; Sozomen, *Eccles. Hist.* 6.34.9. See P. Rousseau, *Basil of Caesarea* 139–44; Miller, *The Birth of the Hospital* 74–5, 85–8; Crislip 103–20.

86. On the date (whether 368, 369, or 370), see Brown, *Poverty and Leadership* 41 and 126 n. 154.

87. Gregory of Nazianzus, Oration 43.61–63.

88. Crislip 105–18.

89. Patlagean, "The Poor" 23.

90. Oration 20, quoted in *DCA* 1:786., s.v. "Hospitals." On Basil as a civic patron who fulfilled the role inherent in classical euergetism, see Finn 222–36, 260.

91. Crislip 8–11.

92. Ibid., 138.

93. Ibid., 40.

94. Ibid., 40–2.

95. He does, however, point out the continuity between them (ibid., 107–8).

96. Brown, *Poverty and Leadership* 74.

97. Nutton, Essay Review 218–21.

98. P. Horden, "The Byzantine Welfare State: Image and Reality," *Bulletin of the Society for the Social History of Medicine* 37 (1985): 9.

99. Ibid.

100. Eustathius's career and influence on Basil remain problematical: see Brown, *Poverty and Leadership* 36–9. Timothy Miller (*The Birth of the Hospital* 74–85) argues that the

hospital originated in the medical charities of Arian churches of Asia Minor. He names four Arians who founded charitable institutions that were protohospitals (one of whom, Sampson, who is usually dated to the sixth century, he argues for moving back to the fourth). Against this view of Arian origins (which is supported by Horden, "Christian Hospital in Late Antiquity" 80–4, 86–7), see Nutton, Essay Review 218–21, and Crislip 128–33.

101. See J. Boswell, *The Kindness of Strangers* 138–79.

102. Ibid., 53–137. Christian parents sometimes abandoned their children too (see 177–9).

103. Brown, *Poverty and Leadership* 35.

104. Horden, "The Byzantine Welfare State: Image and Reality" 9.

105. Crislip 132.

106. Nutton, Essay Review 220. Peter Brown points to a cramped hostel attached to the spacious monastery of St. Martyrius between Jerusalem and Jericho that could have housed sixty or seventy persons (*Poverty and Leadership* 33 and 122 n. 115, citing Y. Magen).

107. Nutton, Essay Review 221.

108. Miller, *The Birth of the Hospital* 26 and 98.

109. Nutton, "From Galen to Alexander" X 10.

110. Nutton, Essay Review 220. Nutton notes that there was already a hospice, "which was itself only a small institution in a substantial city." On the plague, see Segal, *Edessa, the "Blessed City."*

111. Joshua Stylites, *Chron.* 26.28.41–3.

112. Crislip 139–42.

113. Finn 82–8, 263.

114. Jones, *The Later Roman Empire* 1:895–901.

115. Peter Brown lists 40 *xenodocheia* that were founded in Constantinople, together with 59 *xenodocheia*, 45 hospitals, and 22 poorhouses that were founded outside the city, between the fourth and the eighth centuries (*Poverty and Leadership* 122 n. 114, citing K. Mentzou-Meimari). The attempt by bishops to institutionalize the hospital can be seen in the fact that Syriac canons, apparently dating from the fifth century, required that *xenodocheia* be established in each town in the province, where the bishop was to appoint a monk as overseer (Finn 86 and n. 250, citing Vööbus). Not every *xenodocheion*, however, was a hospital in a medical sense.

116. See Jerome's memoir of Fabiola in his letter to Oceanus (Epistle 77), which was written in the year of her death (399). Basil of Caesarea personally supervised the soup kitchen that he founded and nursed lepers, giving them the kiss of peace (Gregory of Nyssa, *Against Eunomius* 1.103; Gregory of Nazianzus, Oration 43.64; Brown, *Poverty and Leadership* 40).

117. "I think," he writes, "that when the poor happened to be neglected and overlooked by the priests, the impious Galileans observed this and devoted themselves to benevolence" (Epistle 49). In another letter Julian speaks of Christians' "benevolence to strangers and their care for the graves of the dead" (Epistle 22). See E. Kislinger, "Kaiser Julian und die (christlichen) Xenodocheia," in *Byzantios: Festschrift für Herbert Hunger zum 70. Geburtstag*, ed. W. Hörandner et al., 171–84; on the motives for Julian's philanthropic program, see Praet, "Explaining the Christianization of the Roman Empire" 104–6.

118. P. Horden, "How Medicalised Were Byzantine Hospitals?" *Medicina e Storia* 10 (2006): 49–50.

119. Ibid., 60.

120. Ibid., 46–7.

121. Ibid., 64, 66–7.

122. P. Horden, "A Non-natural Environment: Medicine without Doctors and the Medieval European Hospital," in *The Medieval Hospital and Medical Practice*, ed. B. S. Bowers, 140–3.

123. See Horden, "How Medicalised Were Byzantine Hospitals?" 69 n. 77 for citations.

124. Crislip 14–5; Keenan, "St. Gregory of Nazianzus and Early Byzantine Medicine" 8–30 and "St. Gregory of Nyssa and the Medical Profession" 150–61.

125. The emphasis of hospitals came to differ according to the region in which they were established. Western European hospitals were more like hospices, while the largest of Byzantine hospitals emphasized secular medicine (J. Henderson, P. Horden, and A. Pastore, eds., *The Impact of Hospitals 300–2000* 21).

126. For the cities in which *spoudaioi* and *philoponoi* were found, see Cabrol and Leclercq, *Dictionnaire*, s.v. "Confréries," 3/2, cols. 2553–8, by H. Leclercq, and S. Pétridès, "Spoudaei et philopones," *Echos d'Orient* 7 (1904): 341–6.

127. See E. Wipszycka, "Les confréries dans la vie religieuse de l'Egypte chretienne," in *Proceedings of the Twelfth International Congress of Papyrology*, ed. D. H. Samuel, 513–5.

128. Timothy Miller calls them "urban ascetics" (*The Birth of the Hospital* 130–1). Although ascetics, they were not a monastic order. Cf. P. Horden, "The Confraternities of Byzantium," in *Voluntary Religion*, ed. W. S. Sheils and D. Wood, 40 n. 71.

129. See Wipszycka, "Les confréries" 515.

130. See Cabrol and Leclercq, *Dictionnaire*, s.v. "Confréries," col. 2553.

131. For a description of the situation of the poor in the late Roman Empire, including many who were diseased and disabled, see Finn 18–26.

132. *Epidemics* 3, case 8, and 1.20. On these cases, see H. N. Couch, "The Hippocratean Patient and His Physician," *TAPA* 65 (1934): 158–60. Some of the locales given may be addresses.

133. Merideth 145–6; Brown, *Poverty and Leadership* 11–3. Brown sees them depicted in early Christian sources as passive and anonymous, needing protection and receiving assistance (13–4).

134. Brown, *Poverty and Leadership* 45–73.

135. Ibid., 16.

136. There is no Greek or Latin word for *family* in the modern sense. The household included, in addition to members of the immediate family, both extended members, such as grandparents, and domestic slaves.

137. Crislip 43.

138. Ibid., 67.

139. Ibid., 45.

140. Ibid., 44–5 and 166 n. 32.

141. See Scobie's list of specific diseases, taken from Celsus's *De medicina*, for which baths are recommended for treatment ("Slums, Sanitation, and Mortality" 425).

142. See, e.g., Lk. 17:12 (lepers); Lk. 8:27 (demoniacs).

143. Scobie, "Slums, Sanitation, and Mortality" 419. It was a common practice to deposit unwanted (exposed) infants and slave gladiators on dung or garbage heaps (ibid.).

144. Homily 30 on 1 Corinthians 5 as quoted by Merideth 149.

145. See Downey, "Who Is My Neighbor?" 8–9, 13.

146. *Spoudaioi* were not nurses in any professional sense. Indeed, nurses did not exist as a separate professional designation, as distinguished from both physicians and lay caregivers, in the ancient world (King, "Using the Past" 7–24). Crislip argues that they emerged as a profession within monasteries in the fourth century (15–6).

147. Miller, *The Birth of the Hospital* 156; cf. 124 and 126.

148. Basil, Epistle 94; Palladius, *Vita Chrys;* Kelly, *Golden Mouth* 119–20. John founded several hospitals, which he kept under his control and for which he nominated the staff, including physicians. On his role in Antioch, see A. Natali, "Eglise et évergétisme à Antioche à la fin du IVe siècle d'après Jean Chrysostome," *Studia Patristica* 17 (1982): 1176–84.

149. Their roles should not be confused with those of, e.g., the *hypourgo*i, who were assistants in Byzantine hospitals.

150. Sozomen (*Eccles. Hist.* 8.23) mentions a wealthy and pious woman who was associated with a *philoponeion* in Constantinople, Nikarete, who personally prepared drugs.

151. Miller, *The Birth of the Hospital* 122.

152. Jerome, Epistle 77.

153. E.g., John of Ephesus (*Lives of Mary and Euphemia*) tells of a woman who sought out the sick in public and private places and took them either to her house or to hospitals.

154. The Council of Chalcedon (451) formally defined the doctrine that in the Incarnation two natures (divine and human) were united inseparably in the person of Christ. This definition was accepted by the Chalcedonians, who became the orthodox party. The Monophysites maintained that Christ had a single divine nature. The division lasted for two centuries in the Eastern church.

155. H. J. Magoulias, "The Lives of the Saints as Sources of Data for the History of Byzantine Medicine in the Sixth and Seventh Centuries," *Byzantinische Zeitschrift* 57 (1964): 135.

156. See Wipszycka, "Les confréries" 511; Horden, "The Confraternities of Byzantium" 40–4.

157. For a conspectus of the papyri, see Wipszycka, "Les confréries" 522–4, who lists all the sources that mention *spoudaioi* and *philoponoi.*

158. P. Iand. 154.

159. Crum, BM 54.

160. See Cabrol and Leclercq, *Dictionnaire,* s.v. "Parabalani," cols. 1575–7.

161. See Horden, "The Confraternities of Byzantium" 41, and Wipszycka, "Les confréries" 515.

162. On the large staffs maintained by the urban churches of the late Roman Empire, see Jones, *The Later Roman Empire* 2:911.

163. Horden, "The Confraternities of Byzantium" 40.

164. Miller (*The Birth of the Hospital* 131–2) thinks that they initially worked in hospitals but were forced to abandon them by the sixth century, when they surrendered their roles in *xenones* to laypeople. There is, however, no evidence that they were ever associated with hospitals and, therefore, no later absence that needs to be explained.

165. Wipszycka, "Les confréries" 519–20.

166. Magoulias, "The Lives of the Saints" 136–7.

167. E.g., by Miller, *The Birth of the Hospital* 131–2. Menas, who was in charge of a *philoponeion* in Alexandria, sought medical aid when ill (see Magoulias, "The Lives of the Saints" 148).

168. Case, *Experience with the Supernatural in Early Christian Times* 229.

169. See E. Thrämer, "Health and Gods of Healing (Greek)," in *ERE* 6:540–3; and idem, "Health and Gods of Healing (Roman)," in *ERE* 6:553–6.

170. See the comprehensive census of 732 sites that comprises Band 2 of Jürgen Riethmüller's detailed description of asclepieia (J. W. Riethmüller, *Asklepios. Heiligtümer und Kulte).*

171. On incubation in pagan and Christian healing, see Hamilton, *Incubation.*

172. See S. V. MacCasland, "Religious Healing in First-Century Palestine," in *Environmental Factors in Christian History*, ed. J. T. MacNeill, M. Spinka, and H. R. Willoughby, 27–34.

173. For the texts and translation of Tablets A and B of the *iamata* from Epidaurus (which date from the second half of the fourth century B.C.), see Edelstein and Edelstein, *Asclepius* 2:221–37. The third and fourth stelae (C and D) can be found in L. R. LiDonnici, *The Epidaurian Miracle Inscriptions: Text, Translation, and Commentary* 116–31.

174. Edelstein and Edelstein, *Asclepius* 2:175–6; H. E. Sigerist, *A History of Medicine* 2:73.

175. See C. A. Behr, *Aelius Aristides and the Sacred Tales*, and idem, "Studies on the Biography of Aelius Aristides," in *ANRW* II. 34, 2 (1994): 1140–233.

176. Edelstein and Edelstein, *Asclepius* 2:173–80.

177. G. Vlastos, "Religion and Medicine in the Cult of Asclepius: A Review Article," *Review of Religion* 13 (1949): 288–90. Very few cases actually involved long-term residence at a shrine.

178. Edelstein and Edelstein, *Asclepius* 132–8.

179. Dodds, *Pagan and Christian* 137; for a discussion of Dodds's view (with which Peter Brown disagrees), see Praet, "Explaining the Christianization of the Roman Empire" 73–5.

180. Paul Johnson, as quoted in Stark 84.

181. Care in hospitals was available not only to Christians but to all persons because of the belief that God's goodness extended to all humans. On this theme in both Gregory of Nazianzus and Gregory of Nyssa, see Holman, *The Hungry Are Dying* 150.

182. Dodds, *Pagan and Christian* 136–7; cf. Veyne, *Bread and Circuses* 23; Hamel, *Poverty and Charity in Roman Palestine* 229–38; Lane Fox, *Pagans and Christians* 324–5.

183. Dodds, *Pagan and Christian* 138; so also Harnack, *Mission and Expansion* 1:181–249.

Chapter 7 • Some Concluding Observations

1. *Contra Celsum* 8.60.

2. Quoted by Brown, "The Saint as Exemplar in Late Antiquity" 11.

3. From an Attic skolion (drinking song) quoted by Plato, *Gorgias* 415E. It is translated by C. M. Bowra, *The Greek Experience* 103–4.

4. *Adv. Mathem.* 11.49. Cited in Edelstein, "Professional Ethics" 357.

5. Quoted by Sextus Empiricus, *Adv. Mathem* 11.50. It is quoted in translation in Edelstein, "The Relation of Ancient Philosophy to Medicine" 358.

6. For the religious implications, see Garland, *The Eye of the Beholder* 63–4. It was a view that pagan ascetics abandoned, however, even before late antiquity.

7. Crislip 39. Associated with the excessive emphasis on health and the body in the second century was the widespread tendency toward hypochondria in the Roman Empire (ibid., 163 n. 3). Several scholars (e.g., Michel Foucault, Ludwig Edelstein, and Judith Perkins) have described this "discourse of a suffering subjectivity" in different ways, and E. R. Dodds and G. W. Bowersock have attributed it to a general cultural anxiety or a reconfiguration of the self (Flemming, *Medicine and the Making of Roman Women* 65 and 71).

8. For the classical belief that a congenitally deformed child was a victim of its parents' iniquity and the resulting anger of the gods, see Garland, *The Eye of the Beholder* 59–61. "It was a strategy which had the incidental benefit of providing a justification for treating such persons as outcasts" (59).

9. Crislip 76–8, 99.

10. Oration 43.63, trans. Browne and Swallow, as quoted in Crislip 119; cf. similar sentiments of Shenoute quoted by Crislip (137).

11. For what follows I am indebted to Holman, "Healing the Social Leper" 294–8.

12. Ibid., 304–9.

13. Crislip attributes the new social role of the sick to monasticism (99), but we have seen it earlier in the charitable concern for the ill that was manifested in the pre-Constantinian churches.

14. H. E. Sigerist, "The Special Position of the Sick," in *Culture, Disease, and Healing*, ed. D. Landy, 391; quoted in Crislip 69.

15. Crislip 39.

16. Ibid., 99.

17. Sigerist, *Civilization and Disease* 69–70; cf. idem, "The Special Position of the Sick" 391–2.

18. On the visitation of the sick in the Jewish community, see Rosner, "Jewish Medicine in the Talmudic Period" 2890–2; cf. Loewenberg, *From Charity to Social Justice* 146.

19. Sigerist, *Civilization and Disease* 69.

20. In fairness to Sigerist, it should be noted that he emphasizes the Christian obligation to care for the sick (ibid., 69–70).

21. Nutton, "Murders and Miracles" VIII 46.

22. Ibid., 45; cf. "From Galen to Alexander" X 5.

23. Gregory of Nyssa, *De pauperibus amandis* 2 (the passage is quoted both in the original and in translation in Holman, "Healing the Social Leper" 294 and n. 55); cf. Gregory Nazianzus, Oration 14.27, quoted in translation in Holman, "Healing the Social Leper" 295.

24. Crislip 26 and 92–9.

25. S. A. Harvey, "Physicians and Ascetics in John of Ephesus: An Expedient Alliance," in *Symposium on Byzantine Medicine*, ed. John Scarborough, 89.

26. Ibid., 92–3.

27. Crislip 27–2, 34–5, 151 n. 36, 158 n. 122, and 163 n. 189. Crislip observes that those monks who eschewed medicine were primarily from Syria (162 n. 175).

28. Merideth 58; cf. 53.

29. Edelstein, "The Distinctive Hellenism of Greek Medicine" in Edelstein 387.

30. Nutton, "Murders and Miracles" VIII 45. Geoffrey Lloyd, too, remarks on this theme (*In the Grip of Disease* 233). Vivian Nutton is right in recognizing this attitude as a novel one that is not found in pagan sources. It grows out of a certain hope of a blessed afterlife together with a willingness to accept suffering as a means of glorifying God, both of which were widely shared beliefs.

31. *De mortalitate,* ch. 9.

32. *De mortalitate,* chs. 15–20, trans. M. H. Mahoney; quoted in Stark 81.

33. *De mortalitate,* chs. 15–20; quoted in Stark 212.

34. *De anima* 30.

35. Crislip 96.

36. Ibid., 97.

37. Stark 77–82.

38. Ibid., 74.

39. *Eccles. Hist.* 7.22, trans. G. A. Williamson.

40. S. A. Harvey, *Asceticism and Society in Crisis: John of Ephesus and the Lives of the Eastern Saints* 4–27, esp. 10–3.

41. P. Brown, *The Body and Society: Men, Women, and Sexual Renunciation in Early Christianity* 222–6 and 235–9.

42. Crislip 74–6 and 178 n. 44.

43. See Gourevitch, *Le Triangle hippocratique* 347–414.

44. Rousselle observes that early Egyptian monks based their regimen on the medical opposition between the wet and the dry and (to a lesser extent) the hot and the cold (see, e.g., Jerome, *Letters* 54.9). Late antique Christian texts "are full of these notions of physiology linked with dietetics" that were common throughout the Mediterranean world (Rousselle, *Porneia* 174–8).

45. T. B. Macaulay, "Life of Lord Bacon," in *Biographical Essays* 117. Macaulay is paraphrasing Francis Bacon in *De augmentis,* bk. 4, ch. 2.

BIBLIOGRAPHY

Achtemeier, Paul J. "Jesus and the Disciples as Miracle Workers in the Apocryphal New Testament." In Fiorenza 1976, 149–86.

Ackerknecht, Erwin H. *A Short History of Medicine.* Rev. ed. Baltimore: Johns Hopkins University Press, 1982.

Agrimi, Jole, and Chiara Crisciani. "Charity and Aid in Medieval Christian Civilization." In Grmek 1998, 170–96.

Aland, Kurt, Matthew Black, Carlo M. Martini, Bruce M. Metzger, and Allen Wikgren, eds. *The Greek New Testament.* 3d ed. (corrected). London: United Bible Societies, 1983.

Alexander, William M. *Demonic Possession in the New Testament: Its Historical, Medical, and Theological Aspects.* 1902. Reprint, Grand Rapids, Mich.: Baker, 1980.

Allan, Nigel. "Hospice to Hospital in the Near East: An Instance of Continuity and Change in Late Antiquity." *BHM* 64 (1990): 446–62.

——. "The Physician in Ancient Israel: His Status and Function." *Medical History* 45 (2001): 377–94.

Allen, P. "The 'Justinianic' Plague." *Byzantion* 49 (1979): 5–20.

Amundsen, Darrel W. "Images of Physicians in Classical Times." *Journal of Popular Culture* 11 (1977): 642–55.

——. "The Physician's Obligation to Prolong Life: A Medical Duty without Classical Roots." *Hastings Center Report* 8, no. 4 (1978): 23–30. Reprinted in Amundsen 1996, 30–49.

——. "Medicine and Faith in Early Christianity," *BHM* 56 (1982): 326–50. Reprinted in Amundsen 1996, 127–57.

——. "Medicine and the Birth of Defective Children: Approaches of the Ancient World." In McMillan et al. 1987, 3–22. Reprinted in Amundsen 1996, 50–69.

——. Review of *Medicine, Miracle, and Magic in New Testament Times* by Howard Clark Kee. *BHM* 63 (1989): 140–4.

——. "Suicide and Early Christian Values." In *Suicide and Euthanasia: Historical and Contemporary Themes,* edited by Baruch A. Brody, 77–153. Dordrecht: Kluwer Academic Publishers, 1989. Reprinted in Amundsen 1996, 70–126.

——. "History of Medical Ethics: The Ancient Near East." In Reich 1995, 3:1440–5.

——. "Tatian's 'Rejection' of Medicine in the Second Century." In van der Eijk et al. 1995, 2:377–92. Reprinted in Amundsen 1996, 158–74.

——. "Body, Soul and Physician." In Amundsen 1996, 1–29.

——. *Medicine, Society, and Faith in the Ancient and Medieval Worlds.* Baltimore: Johns Hopkins University Press, 1996.

——. "The Discourses of Early Christian Medical Ethics." In *The Cambridge World History of Medical Ethics,* edited by Robert Baker and Lawrence McCullough, 202 ff. New York: Cambridge University Press, 2009.

Amundsen, Darrel W., and Gary B. Ferngren. "Medicine and Religion: Pre-Christian Antiquity" and "Medicine and Religion: Early Christianity through the Middle Ages." In *Health/Medicine and the Faith Traditions: An Inquiry into Religion and Medicine,* edited by Martin E. Marty and Kenneth L. Vaux, 53–92 and 93–131. Philadelphia: Fortress Press, 1982.

——. "Virtue and Medicine from Early Christianity through the Sixteenth Century." In Shelp 1985, 23–61.

——. "The Early Christian Tradition." In *Caring and Curing: Health and Medicine in the Western Religious Traditions,* edited by Ronald L. Numbers and Darrel W. Amundsen, 40–64. New York: Macmillan, 1986.

——. "The Perception of Disease and Disease Causality in the New Testament." In *ANRW* II. 37, 3 (1996): 2934–56.

——. "The Healing Miracles of the Gospels: Problems and Methods." In *ANRW* II. 26, 3 (forthcoming).

Angus, S. *The Religious Quests of the Graeco-Roman World: A Study in the Historical Background of Early Christianity.* New York: Charles Scribner's Sons, 1929.

Arbesmann, R. "The Concept of 'Christus Medicus' in St. Augustine." *Traditio* 10 (1954): 1–28.

Arndt, William F., and F. Wilbur Gingrich. *A Greek-English Lexicon of the New Testament and Other Early Christian Literature.* 4th ed. Chicago: University of Chicago Press, 1957.

Arnobius of Sicca. *The Case against the Pagans.* 2 vols. Newly translated and annotated by George E. McCracken. New York: Newman Press, 1949.

Augustine. *Sermons.* In *The Works of St. Augustine: A Translation for the 21st Century,* edited by John E. Rotelle, OSA, translated by Edmund Hill, OP. 9 vols. Brooklyn, N.Y.: New City Press, 1991.

Aune, David. "Magic in Early Christianity." In *ANRW* II. 23, 2 (1981): 1507–57.

——. *Prophecy in Early Christianity and the Ancient Mediterranean World.* Grand Rapids, Mich.: Eerdmans, 1983.

Avalos, Hector. *Illness and Health Care in the Ancient Near East: The Role of the Temple in Greece, Mesopotamia, and Israel.* Harvard Semitic Monographs 54. Atlanta: Scholars Press, 1995.

——. *Health Care and the Rise of Christianity.* Peabody, Mass.: Hendrickson, 1999.

Babcork, W. S. "MacMullen on Conversion: A Response." *Second Century* 5 (1985–86): 82–9.

Bachmann, U. "Medizinisches in den Schriften des griechischen Kirchenvators Johannes Chrysostomos." Diss. med., University of Düsseldorf, 1984.

Bailey, Michael D. *Magic and Superstition in Europe: A Concise History from Antiquity to the Present.* Lanham, Md.: Rowman and Littlefield, 2007.

Baldwin, Barry. "The Career and Works of Scribonius Largus." *Rheinisches Museum* 135 (1992): 74–82.

Barb, A. A. "The Survival of Magic Arts." In Momigliano 1963, 100–25.

Barker, E. *From Alexander to Constantine.* Oxford: Clarendon Press, 1956.

Barnes, Timothy D. "The Chronology of Montanism." *Journal of Theological Studies*, n.s., 20 (1970): 403–8.

——. *Tertullian: A Historical and Literary Study.* 1971. Oxford: Clarendon Press, 1985.

Baumgarten, Joseph M. "The 4Q Zadokite Fragments on Skin Disease." *Journal of Jewish Studies* 41 (1990): 153–65.

Baziotopoulou-Valavani, E. "A Mass Burial from the Cemetery of Kerameikos." In *Excavating Classical Culture: Recent Archaeological Discoveries in Greece*, Studies in Classical Archaeology 1, edited by Maria Stamatopoulou and Marina Yeroulanou, 187–201. Oxford: Archaeopress, 2002.

Beato, L. *Teologia dell malattia in S. Ambrogio.* Turin, 1986.

Behr, Charles A. *Aelius Aristides and the Sacred Tales.* Amsterdam: Hakkert, 1968.

——. "Studies on the Biography of Aelius Aristides." In *ANRW* II. 34, 2 (1994): 1140–1223.

Belkin, S. *In His Image: The Jewish Philosophy of Man as Expressed in Rabbinic Tradition.* London: Abelard-Schuman, 1960.

Bell, H. I. "Philanthropy in the Papyri of the Roman Period." *Hommages à Joseph Bidez et à Franz Cumont,* Collection Latomus II, Brussels (1948): 31–7.

Bellemare, P. M. "The Hippocratic Oath: Edelstein Revisited." In *Healing in Religion and Society from Hippocrates to the Puritans: Selected Studies*, edited by J. K. Coyle and S. C. Muir, 1–64. Lewiston, N.Y.: Edwin Mellen Press, 1999.

Benko, Stephen. "Early Christian Magical Practices." *SBL 1982 Seminar Papers*, 9–14.

——. *Pagan Rome and the Early Christians.* Bloomington: Indiana University Press, 1984.

Berardino, Angelo Di, ed. *Encyclopedia of the Early Church.* Translated by Adrian Walford. 2 vols. New York: Oxford University Press, 1992. S.v. "Arnobius of Sicca," 1:82, by P. Sinalesco.

Berkowitz, Lucille. *Index Arnobianus.* Hildesheim: Georg Olms, 1967.

Bernard, L. W. "Athenagoras: De Resurrectione. The Background and Theology of a Second Century Treatise on the Resurrection." *Studia Theologica* 30 (1976): 1–42.

Betz, Hans Dieter, ed. *The Greek Magical Papyri in Translation including the Demotic Spells.* Chicago: University of Chicago Press, 1986.

Böcher, Otto. *Dämonenfurcht und Dämonenabwehr: Ein Beitrag zur Vorgeschichte der christlichen Taufe.* Stuttgart: W. Kohlhammer, 1970.

——. *Christus Exorcista. Dämonismus und Taufe im Neuen Testament.* Stuttgart: W. Kohlhammer, 1972.

——. *Das neue Testament und die dämonischen Mächte.* Stuttgart: W. Kohlhammer, 1972.

Bolkestein, Hendrick. *Wohltätigkeit und Armenpflege in vorchristlichen Altertum.* 1939. Reprint, Groningen: Bouman Boekhuis, 1967.

Bonnechere, Pierre. *Trophonios de Lébadée: Cultes et mythes d'une cité béotienne au miroir de la mentalité antique.* Leiden: Brill, 2003.

Borg, Marcus. *Jesus: A New Vision.* San Francisco: Harper, 1987.

Bostock, D. G. "Medical Theory and Theology in Origen." In *Origeniana Tertia: The Third International Colloquium for Origen Studies*, 191–9. Manchester: University of Manchester, 1981.

Boswell, John. *The Kindness of Strangers: The Abandonment of Children in Western Europe from Late Antiquity to the Renaissance.* New York: Pantheon, 1988.

Botterweck, G. Johannes, and Helmer Ringgren, eds. *Theological Dictionary of the Old Testament.* Translated by J. T. Willis. 15 vols. Grand Rapids, Mich.: Eerdmans, 1974–.

Bottomley, Frank. *Attitudes to the Body in Western Christendom.* London: Lepus, 1979.

Bowers, Barbara S., ed. *The Medieval Hospital and Medical Practice.* Aldershot, England: Ashgate, 2007.

Bowersock, G. W. *Greek Sophists in the Roman Empire.* Oxford: Clarendon Press, 1969.

Bowra, C. M. *The Greek Experience.* 1957. New York: Praeger, 1969.

Brain, Peter. "Galen on the Ideal of the Physician." *South African Medical Journal* 52 (1977): 936–8.

Breitenbach, A. "Wer Christlich Lebt, Lebt Gesund." In *Jahrbuch für Antike und Christentum*, 24–49. Münster Westfalen: Aschendorff Verlag, 2002.

Brenk, Frederick E., SJ. "In the Light of the Moon: Demonology in the Early Imperial Period." In *ANRW* II. 16, 3 (1986): 2068–145.

Brock, R. "Sickness in the Body Politic: Medical Imagery in the Greek Polis." In Hope and Marshall 2000, 24–34.

Bromiley, Geoffrey W. "Image of God." In *The International Standard Bible Encyclopedia*, edited by Geoffrey W. Bromiley, rev. ed., 2:803–5. Grand Rapids, Mich.: Eerdmans, 1979–88.

Brooke, John Hedley, ed. *Science and Religion: Some Historical Perspectives.* Cambridge: Cambridge University Press, 1991.

Brown, Colin, ed. *The New International Dictionary of New Testament Theology.* 3 vols. Grand Rapids, Mich.: Zondervan, 1976. S.v. "Miracle," 2:626–35, by O. Hofius. S.v. "Weakness et al.," 3:993–1000, by H.-G. Link and R. K. Harrison. S.v. "Fight et al.," 1:644–52, by W. Günther et al.

Brown, Peter. "Approaches to the Religious Crisis of the Third Century A.D." *English Historical Review* 83 (1968): 542–58. Reprinted in Brown 1972, 74–93.

——. *Augustine of Hippo: A Biography.* Berkeley: University of California Press, 1969.

——. "Sorcery, Demons, and the Rise of Christianity from Late Antiquity into the Middle Ages." In *Witchcraft Confessions and Accusations*, edited by Mary Douglas, 17–45. 1970. Reprint, London: Routledge, 2004. Reprinted in Brown 1972, 119–46.

——. "The Rise and Function of the Holy Man in Late Antiquity." *Journal of Roman Studies* 61 (1971): 80–101. Reprinted in Brown 1982 (*Society and the Holy*), 103–52.

——. *The World of Late Antiquity, A.D. 150–750.* New York: Harcourt Brace Jovanovich, 1971.

——. *Religion and Society in the Age of Saint Augustine.* New York: Harper and Row, 1972.

——. *The Making of Late Antiquity.* Cambridge: Harvard University Press, 1978.

——. *The Cult of the Saints: Its Rise and Function in Latin Christianity.* Chicago: University of Chicago Press, 1982.

——. *Society and the Holy in Late Antiquity.* Berkeley: University of California Press, 1982.

——. "The Saint as Exemplar in Late Antiquity." *Representations* 2 (1983): 1–25.

——. *The Body and Society: Men, Women and Sexual Renunciation in Early Christianity.* New York: Columbia University Press, 1988.

——. *Power and Persuasion in Late Antiquity: Towards a Christian Empire.* Madison: University of Wisconsin Press, 1992.

——. *Authority and the Sacred.* Cambridge: Cambridge University Press, 1995.

——. *Poverty and Leadership in the Later Roman Empire.* Hanover, N.H.: University Press of New England, 2002.

Browning, Robert. "The 'Low Level' Saint's Life in the Early Byzantine World." In *The Byzantine Saint,* edited by S. Hackel, 117–27. London: Fellowship of St. Albans and St. Sergius, 1991.

Brunt, P. A. "Aspects of the Social Thought of Dio Chrysostom and the Stoics." *Proceedings of the Cambridge Philological Society* N.W. 19 (1973): 26–34.

——. "The Bubble of the Second Sophistic." *Bulletin of the Institute of Classical Studies* 40 (1994): 25–52.

Burckhardt, Jacob. *The Greeks and Greek Civilization.* Translated by Sheila Stern, edited by Oswyn Murray. New York: St. Martin's Press, 1998.

Bynum, Caroline Walker. *The Resurrection of the Body in Western Christianity, 200–1336.* New York: Columbia University Press, 1995.

Bynum, W. F., and Roy Porter, eds. *Companion Encyclopedia of the History of Medicine.* 2 vols. London: Routledge, 1993.

Cabrol, Fernand, and H. Leclercq, eds. *Dictionnaire d'archéologie chrétienne et de liturgie.* 15 volumes. Paris: Letouzeyet et Ané, 1907–53. S.v. "Confréries," 3/2, cols. 2553–60, by H. Leclercq. S.v. "Parabalani," 13/2, cols. 1574–8, by H. Leclercq.

Cadbury, Henry Joel. *The Style and Literary Method of Luke.* 1919–20. Reprint, New York: Kraus, 1969.

Cameron, Averil. *Christianity and the Rhetoric of Empire: The Development of Christian Discourse.* Berkeley: University of California Press, 1991.

Case, Shirley Jackson. "The Art of Healing in Early Christian Times." *Journal of Religion* 3 (1923): 238–55.

——. *Experience with the Supernatural in Early Christian Times.* New York: Century, 1929.

Carrick, P. *Medical Ethics in Antiquity: Philosophical Perspectives on Abortion and Euthanasia.* Dordrecht: Reidel, 1985.

Castelli, Elizabeth A. "Mortifying the Body, Curing the Soul: Beyond Ascetic Dualism in *The Life of Saint Syncletica.*" *Differences* 4 (1992): 134–53.

——. "Gender, Theory, and *The Rise of Christianity*: A Response to Rodney Stark." *Journal of Early Christian Studies* 6 (1998): 227–57.

Castrén, P., ed. *Ancient and Popular Healing.* Athens: Finnish Institute in Athens, 1989.

Cavarnos, John P. "Relation of the Body and Soul in the Thought of Gregory of Nyssa." In *Gregor von Nyssa und die Philosophie: Zweites Internationales Kolloquium über Gregor von Nyssa,* edited by Heinrich Dörrie et al., 61–78. Leiden: Brill, 1976.

Charlesworth, James H. *The Discovery of a Dead Sea Scroll (4Q Therapeia): Its Importance in the History of Medicine and Jesus Research.* Lubbock: Texas Tech University, 1985.

——. "A Misunderstood Recently Published Dead Sea Scroll." *Explorations* 1, no. 2 (1994): 2.

Clark, Gillian. *Women in Late Antiquity.* Oxford: Clarendon Press, 1994.

Clarke, M. L. *The Roman Mind.* New York: Norton, 1968.

Collingwood, R. G. *The Idea of History.* 1946. London: Oxford University Press, 1961.

Colson, Jean. *La Fonction diaconale aux origines de l'Église.* Paris: Desclée de Brouwer, 1960.

Connery, John, SJ. *Abortion: The Development of the Roman Catholic Perspective.* Chicago: Loyola University Press, 1977.

Conrad, Lawrence I., and Dominik Wujastyk, eds. *Contagion: Perspectives from Pre-Modern Societies.* Aldershot, England: Ashgate, 2000.

Constantelos, Demetrios J. *Byzantine Philanthropy and Social Welfare.* 2d ed. New Rochelle, N.Y.: Aristide D. Caratzas, 1991.

Cooke, Bernard J. *Ministry to Word and Sacraments: History and Theology.* Philadelphia: Fortress, 1976.

Cordes, Peter Iatros. *Das Bild des Ärztes in der Griechischen Literatur von Homer bis Aristoteles.* Stuttgart: Franz Steiner, 1994.

Cortés, Juan B., and Florence M. Gatti. *The Case against Possessions and Exorcisms: A Historical, Biblical, and Psychological Analysis of Demons, Devils, and Demoniacs.* New York: Vantage Press, 1975.

Couch, Herbert Newell. "The Hippocratean Patient and His Physician." *TAPA* 65 (1934): 138–62.

Courtès, J. "Augustin et la medicine." In *Augustinus Magister,* 1:43–51. Paris: Études Augustiniennes, 1954–55.

Cranfield, C. E. B. "Diakonia in the New Testament." In McCord and Parker 1966, 37–48.

Crislip, Andrew T. *From Monastery to Hospital: Christian Monasticism and the Transformation of Health Care in Late Antiquity.* Ann Arbor: University of Michigan Press, 2005.

Crossan, John Dominic. *The Historical Jesus: The Life of a Mediterranean Jewish Peasant.* San Francisco: Harper, 1991.

Cruttwell, Charles Thomas. *A Literary History of Early Christianity.* 2 vols. 1893. Reprint, New York: AMS Press, 1971.

Csepregi, Ildikó. "The Compositional History of Greek Christian Incubation Miracle Collections: Saint Thecla, Saint Cosmas and Damian, Saint Cyrus and John, Saint Artemios." Ph.D. diss., Central European University, Budapest, 2007.

Cuesta, J. *La antropología y la medicina pastoral de San Gregorio de Nysa.* Madrid: Consejo Superior de Investigaciones Cientificas, 1946.

Cuming, Goeffrey J. *Hippolytus: A Text for Students with Introduction, Translation, Commentary and Notes.* Bramcote, Notts, England: Grove Books, 1979.

Davidson, W. L. "Image of God." In *ERE* 7:160–4.

Davies, J. G. "Deacons, Deaconesses and the Minor Orders in the Patristic Period." *Journal of Ecclesiastical History* 14 (1963): 1–15.

Davies, Stevan L. *Jesus the Healer: Possession, Trance, and the Origins of Christianity.* New York: Continuum, 1995.

Dawe, V. G. "The Attitude of the Ancient Church toward Sickness and Health." Ph.D. diss., Boston University School of Theology, 1955.

Déaut, R. le. "Philanthropia dans la littérature grecque Jusqu'au Nouveau Testament." In *Studie Testi: Mélanges Eugène Tisserant,* 1:255–94. Vatican City: Biblioteca Apostolica Vaticana, 1964.

Deichgräber, K. *Professio medici: Zum Vorwort des Scribonius Largus.* Abhandlungen der Akademie der Wissenschaften und der Literature, no. 9. Mainz: Steiner, 1950.

——. *Der hippokratische Eid.* Stuttgart: Hippokrates Verlag, 1983.

De Lacy, Phillip, ed. and trans. *Galen: On the Doctrines of Hippocrates and Plato [De placitis Hippocratis et Platonis]. Corpus Medicorum Graecorum* 5.4.1 and 5.4.2. 2 pts. Pt. 1 (bks. 1–5), 3rd ed. Pt. 2 (bks. 6–9), 2nd ed. Berlin: Akademie-Verlag, 1984.

den Boer, W. *Private Morality in Greece and Rome: Some Historical Aspects.* Leiden: Brill, 1979.

deSilva, David A. *Introducing the Apocrypha: Message, Context, and Significance.* Grand Rapids, Mich.: Baker Academic, 2002.

Dickie, M. *Magic and Magicians in the Greco-Roman World.* London: Routledge, 2001.

Dillon, M. P. J. "The Didactic Nature of the Epidaurian *iamata.*" *Zeitschrift für Papyrologie und Epigraphik* 101 (1994): 239–60.

——. *Pilgrims and Pilgrimage in Ancient Greece.* London: Routledge, 1997.

Dinkler, E. *Christus und Asklepios: Zum Christustypus der polychromen Platten im Museo Nazionale Romano.* Sitzungsberichte der Heidelberger Akademie der Wissenschaften, Philosophisch-historische Klasse 1980 (2). Heidelberg, Winter, 1980.

D'Irsay, Stephen. "Patristic Medicine." *Annals of Medical History* 9 (1927): 364–78.

——. "Christian Medicine and Science in the Third Century." *Journal of Religion* 10 (1930): 515–44.

Dodds, E. R. *The Greeks and the Irrational.* Berkeley: University of California Press, 1964.

——. *Pagan and Christian in an Age of Anxiety: Some Aspects of Religious Experience from Marcus Aurelius to Constantine.* Cambridge: Cambridge University Press, 1968.

Dölger, F. J. "Das Lebensrecht des ungeborenen Kindes und die Fruchtabtreibung in der Bewertung der heidnischen und christlichen Antique." *Antike und Christentum* 4 (1934): 1–61.

——. "Der Einfluss des Origenes auf die Beurteilung der Epilepsie und Mondsucht im christlichen Altertum." *Antike und Christentum* 4 (1934): 95–109.

Dols, Michael W. *Majnūn: The Madman in Medieval Islamic Society.* Edited by D. E. Immisch. Oxford: Clarendon Press, 1992.

Dörnemann, Michael. "Medizinale Inhalte in der Theologie des Origenes." In Schulze and Ihm 2002, 9–39.

——. *Krankheit und Heilung in der Theologie der frühen Kirchenväter.* Studien und Texte zu Antike und Christentum 20. Tübingen: Mohr Siebeck, 2003.

Dover, K. J. *Greek Popular Morality in the Time of Plato and Aristotle.* Oxford: Basil Blackwell, 1974.

Downey, Glanville. "Philanthropia in Religion and Statecraft in the Fourth Century after Christ." *Historia* 4 (1955): 199–208.

——. "Who Is My Neighbor? The Greek and Roman Answer." *Anglican Theological Review* 47 (1965): 3–15.

Dumeige, Gervais, SJ. "Le Christ médecin dans la littérature crétienne des premiers siècles." *Rivista di archeologia cristiana* 47 (1972): 115–41.

Duncan-Jones, R. P. "The Impact of the Antonine Plague." *Journal of Roman Archaeology* 9 (1996): 108–36.

Dunn, James D. G., and Graham H. Twelftree. "Demon Possession and Exorcism in the New Testament." *Churchman* 94 (1980): 210–25.

Dzielska, Maria. *Apollonius of Tyana in Legend and History.* Translated by Piotr Pieńkowski. Rome: "L'Erma" di Bretschneider, 1986.

Edelstein, Emma J., and Ludwig Edelstein. *Asclepius: A Collection and Interpretation of the Testimonies.* With a new introduction by Gary B. Ferngren. 2 vols. in 1. Reprint, Baltimore: Johns Hopkins University Press, 1998. Originally published in 1945.

Edelstein, Ludwig. *The Hippocratic Oath: Text, Translation, and Interpretation.* Supplements to the Bulletin of the History of Medicine, no. 1. Baltimore: Johns Hopkins Press, 1943. Reprinted in Edelstein 1967, 3–63.

——. *Ancient Medicine: Selected Papers of Ludwig Edelstein.* Edited by Owsei Temkin and C. Lilian Temkin. Baltimore: Johns Hopkins Press, 1967.

——. "The Distinctive Hellenism of Greek Medicine." In Edelstein 1967, 369–97.

——. "Greek Medicine in Its Relation to Religion and Magic." In Edelstein 1967, 205–46.

——. "The Professional Ethics of the Greek Physician." In Edelstein 1967, 319–48.

——. "The Relation of Ancient Philosophy to Medicine." In Edelstein 1967, 349–66.

Ehrman, Bart D. *Lost Christianities: The Battles for Scripture and the Faiths We Never Knew.* New York: Oxford University Press, 2003.

Eijkenboom, Petrus Cornelis Josephus. *Het Christus-Medicusmotief in de preken van Sint Augustinus* (Christus Medicus in the Sermons of St. Augustine). Assen, Netherlands: Van Gorcum, 1960.

Eitrem, S. *Some Notes on the Demonology in the New Testament.* Oslo: A. W. Brøgger, 1950.

Empereur, James L., S.J. *Prophetic Anointing: God's Call to the Sick, the Elderly, and the Dying.* Wilmington, Del.: Michael Glazier, 1982.

Entralgo, P. Lain. *Doctor and Patient.* Translated by F. Partridge. New York: McGraw-Hill, 1969.

Etienne, R. "La conscience médicale antique et la vie des enfants." *Enfant et Societé* (1973): 15–46.

Eusebius. *The History of the Church.* Translated by G. A. Williamson. Harmondsworth, England: Penguin, 1965.

Faraone, Christopher A., and Dirk Obbink, eds. *Magika Hiera: Ancient Greek Magic and Religion.* New York: Oxford University Press, 1991.

Fenner, F. *Die Krankheit im Neuen Testament.* Leipzig: J. C. Hinrichs'sche Verlagsbuchhandlung, 1930.

Ferguson, Everett. "The Demons According to Justin Martyr." In *The Man of the Messianic Reign and Other Essays,* 103–12. Witchita Falls, Tex.: Western Christian Foundation, 1980.

——. *Demonology of the Early Christian World.* Symposium Series vol. 12. New York: Edwin Mellen Press, 1984.

——. *Backgrounds of Early Christianity.* Grand Rapids, Mich.: Eerdmans, 1987.

——, ed. *Encyclopedia of Early Christianity.* New York: Garland, 1990.

Ferguson, John. *Moral Values in the Ancient World.* London: Methuen, 1958.

Fernández, Samuel. *Cristo médico según Origenes: La actividad médica como metáfora de la acción divina.* Rome: Institutum Patristicum Augustinianum, 1999.

Ferngren, Gary B. "A Roman Declamation on Vivisection." *Transactions and Studies of the College of Physicians of Philadelphia,* ser. 4, 4 (1982): 272–90.

——. "The *Imago Dei* and the Sanctity of Life: The Origins of an Idea." In McMillan et al. 1987, 23–45.

——. Review of John M. Riddle, *Eve's Herbs: A History of Contraception and Abortion in the West. New England Journal of Medicine* 337 (November 6, 1997): 1398.

——. "Medicine and Compassion in Early Christianity." *Theology Digest* 46, no. 4 (1999): 1–12.

——, gen. ed. *The History of Science and Religion in the Western Tradition: An Encyclopedia.* New York: Garland, 2000.

——. Review of *Healing in the New Testament* by John J. Pilch. *BHM* 78 (2004): 468–9.

Ferngren, Gary B., and Darrel W. Amundsen. "Virtue and Health/Medicine in Pre-Christian Antiquity." In Shelp 1985, 3–22.

——. "Medicine and Christianity in the Roman Empire: Compatibilities and Tensions." In *ANRW* II. 37, 3 (1996): 2957–80.

Ferngren, Gary B., and Karl-Heinz Leven. "Médecine aux premiers siècles du christianisme." *Lettre d'Informations, Centre Jean-Palerne* 26 (1995): 2–22.

Festugière, A.-J. "Epidémies 'hippocratiques' et épidémies démoniaques." *Wiener Studien* 79 (1966): 157–64.

Fichtner, Gerhard. "Christus als Artzt: Ursprünge und Wirkungen eines Motivs." *Frühmittelalterliche Studien* 16 (1982): 1–18.

Finley, M. I. *Ancient Slavery and Modern Ideology.* New York: Viking Press, 1980.

Finn, Richard, OP. *Almsgiving in the Later Roman Empire: Christian Promotion and Practice (313–450).* Oxford: Oxford University Press, 2006.

Fiorenza, Elisabeth Schüssler, ed. *Aspects of Religious Propaganda in Judaism and Early Christianity.* Notre Dame, Ind.: University of Notre Dame Press, 1976.

Fitzgerald, William A. "Medical Men: Canonized Saints." *BHM* 22 (1948): 635–46.

Flashar, Hellmut. *Melancholie und Melancholiker in den medizinschen Theorien der Antike.* Berlin: Walter de Gruyter, 1966.

Flashar, Hellmut, and J. Jouanna, eds. *Médecine et morale dans l'antiquité.* Vandoeuvres, Switzerland: Fondation Hardt, 1997.

Flemming, Rebecca. *Medicine and the Making of Roman Women: Gender, Nature, and Authority from Celsus to Galen.* Oxford: Oxford University Press, 2000.

Flint, Valerie I. J. *The Rise of Magic in Early Medieval Europe.* Princeton, N.J.: Princeton University Press, 1991.

——. "The Demonisation of Magic and Sorcery in Late Antiquity: Christian Redefinitions of Pagan Religions." In *Witchcraft and Magic in Europe: Ancient Greece and Rome,* by V. Flint, R. Gordon, G. Luck, and D. Ogden, 277–348. London: Athlone Press, 1999.

Foucault, Michel. *The Birth of the Clinic: An Archaeology of Medical Perception.* Translated by A. M. Sheridan Smith. New York: Pantheon, 1973.

——. *The Care of the Self.* Vol. 3 of *The History of Sexuality.* Translated by R. Hurley. New York: Pantheon, 1986.

Frankfurter, David. "Amuletic Invocations of Christ for Health and Fortune." In *Religions of Late Antiquity in Practice,* edited by R. Valantasis, 340–3. Princeton, N.J.: Princeton University Press, 2000.

Frede, Michael. "Philosophy and Medicine in Antiquity." In *Human Nature and Natural Knowledge,* edited by A. Donagan, A. N. Perovich Jr., and M. V. Wedin, 211–32. Dordrecht: Reidel, 1986.

Frend, W. H. C. *Martyrdom and Persecution in the Early Church.* New York: New York University Press, 1967.

——. *The Rise of Christianity.* Philadelphia: Fortress, 1984.

Frensch, I. "Die Krankheitsauffassung des Basilius des Großen." Ph.D. diss., Freiburg i. Br., 1965.

Frings, Hermann-Josef. "Medizin und Arzt bei den griechischen Kirchernvätern bis Chrysostomos." Ph.D. diss., University of Bonn, 1959.

Frost, E. *Christian Healing.* London: Mowbray, 1949.

Gager, John G. *Kingdom and Community.* Englewood, N. J.: Prentice Hall, 1975.

——. "Body-Symbols and Social Reality: Resurrection, Incarnation and Asceticism in Early Christianity." *Religion* 12 (1982): 345–64.

——. "Introduction: The Dodds Hypothesis." In *Pagan and Christian Anxiety: A Response to E. R. Dodds,* edited by Robert C. Smith and John Lounibos, 1–11. Lanham, Md.: University Press of America, 1984.

——, ed. *Curse Tablets and Binding Spells from the Ancient World.* New York: Oxford University Press, 1992.

Galvão-Sobrinho, Carlos R. "Hippocratic Ideals, Medical Ethics, and the Practice of Medicine in the Early Middle Ages: The Legacy of the Hippocratic Oath." *JHM* 51 (1996): 438–55.

Garland, Robert. *The Eye of the Beholder: Deformity and Disability in the Graeco-Roman World.* Ithaca, N.Y.: Cornell University Press, 1995.

Garnsey, Peter. *Famine and Food Supply in the Graeco-Roman World: Responses to Risk and Crisis.* Cambridge: Cambridge University Press, 1988.

Garnsey, Peter, and Caroline Humfress. *The Evolution of the Late Antique World.* Cambridge: Orchard Academic, 2001.

Garrison, Roman. *Redemptive Almsgiving in Early Christianity.* Sheffield, England: JSOT Press, 1993.

Garzya, Antonio. "Science et conscience dans la pratique médicale de l'Antiquité tardive et byzantine." In Flashar and Jouanna 1997, 337–59.

Gask, George E., and John Todd. "The Origin of Hospitals." In Underwood 1953, 1:122–30.

Gero, Stephen. "Montanus and Montanism According to a Medieval Syriac Source." *Journal of Theological Studies,* n.s., 28 (1977): 520–4.

——. "Galen on Jews and Christians: A Reappraisal of the Arabic Evidence." *Orientalia Christiana* 56 (1990): 371–411.

Gibbon, Edward. *The History of the Decline and Fall of the Roman Empire.* Edited by J. B. Bury. 7 vols. 1911. Reprint, New York: AMS Press, 1974.

Giordani, Igino. *The Social Message of the Early Church Fathers.* Translated by A. Zizzamia. Boston: St. Paul Editions, 1977.

Goguel, Maurice. *The Primitive Church.* Translated by H. C. Snape. London: George Allen and Unwin, 1963.

Goldin, Judah. "The Magic of Magic and Superstition." In Fiorenza 1976, 115–47.

Good, Byron J. *Medicine, Rationality, and Experience: An Anthropological Perspective.* Cambridge: Cambridge University Press, 1994.

Gourevitch, Danielle. *Le Triangle hippocratique dans le monde gréco-romain: Le Malade, sa maladie et son médecin.* Paris: École française de Rome, 1984.

——. "The Paths of Knowledge: Medicine in the Roman World." In Grmek 1998, 104–38.

Graf, Fritz. "Prayer in Magic and Religious Ritual." In *Magika Hiera: Ancient Greek Magic and Religion,* edited by Christopher A. Faraone and Dirk Obbink, 188–213. New York: Oxford University Press, 1991.

——. *Magic in the Ancient World.* Translated by Franklin Philip. Cambridge: Harvard University Press, 1997.

Grant, Robert M. *Miracle and Natural Law in Graeco-Roman and Early Christian Thought.* Amsterdam: North-Holland, 1952.

——. *A Historical Introduction to the New Testament.* New York: Simon and Schuster, 1972.

——. "The Christian Population of the Roman Empire." In Grant 1977/8, 1–12.

——. *Early Christianity and Society: Seven Studies.* London: Collins, 1977/8.

——. "The Organization of Alms." In Grant 1977/8, 124–45.

——. "Paul, Galen and Origen." *Journal of Theological Studies* 34 (1983): 533–6.

——. *Greek Apologists of the Second Century.* Philadelphia: Westminster Press, 1988.

Green, Michael. *Evangelism in the Early Church.* Grand Rapids, Mich.: Eerdmans, 1970.

Greer, Rowan A. *The Fear of Freedom: A Study of Miracles in the Roman Imperial Church.* University Park: Pennsylvania State University Press, 1989.

Griffith, S. H. "Ephraem, the Deacon of Edessa, and the Church of the Empire." In Halton and Williman 1986, 22–52.

Grmek, Mirko D. *Diseases in the Ancient Greek World.* Translated by Mireille Muellner and Leonard Muellner. Baltimore: Johns Hopkins University Press, 1989.

——, ed. *Western Medical Thought from Antiquity to the Middle Ages.* Cambridge: Harvard University Press, 1998.

Gusmer, Charles W. *And You Visited Me: Sacramental Ministry to the Sick and the Dying.* New York: Pueblo, 1984.

Guthrie, Donald. *New Testament Introduction.* 3d ed., rev. Downers Grove, Ill.: InterVarsity Press, 1970.

Halton, T., and J. P. Williman, eds. *Diakonia: Studies in Honor of Robert T. Meyer.* Washington, D.C.: Catholic University of America Press, 1986.

Hamel, Gildas. *Poverty and Charity in Roman Palestine, First Three Centuries C.E.* Near Eastern Studies 23. Berkeley: University of California Press, 1990.

Hamilton, J. S. "Scribonius Largus on the Medical Profession." *BHM* 60 (1986): 209–16.

Hamilton, Mary. *Incubation: Or, The Cure of Disease in Pagan Temples and Christian Churches.* London: Simpkin, Marshall, Hamilton, Kent, 1906.

Hands, A. R. *Charities and Social Aid in Greece and Rome.* Ithaca, N.Y.: Cornell University Press, 1968.

Hankinson, R. J. "Galen's Theory of Causation." In *ANRW* II. 37, 2 (1994): 1757–74.

Hankoff, L. D. "Religious Healing in First-Century Christianity." *Journal of Psychohistory* 19, no. 4 (1992): 387–407.

Hardon, John A. "The Miracle Narratives in the Acts of the Apostles." *CBQ* 16 (1954): 303–18.

Harig, Georg. "Zum Problem 'Krankenhaus' in der Antike." *Klio* 53 (1971): 179–95.

Harig, Georg, and Jutta Kollesch. "Arzt, Kranker und Krankenpflege in der Griechisch-Römischen Antike und im Byzantinischen Mittelalter." *Helikon* 13/14 (1973): 256–92.

Harnack, Adolf. "Medicinisches aus der ältesten Kirchengeschichte." In *Texte Untersuchungen zur Geschichte der altchristlichen Literatur,* vol. 8, pt. 4, 37–152. Leipzig: J. C. Hinrich, 1892.

———. "The Gospel of the Saviour and of Salvation." In Harnack 1904, 1:121–51.

———. *The Mission and Expansion of Christianity in the First Three Centuries.* Translated and edited by James Moffatt. 3 vols. New York: G. P. Putnam's Sons, 1904.

Harris, Bruce F. "The Idea of Mercy and Its Graeco-Roman Context." In O'Brien and Peterson 1986, 89–105.

Harris, W. V. "Child-Exposure in the Roman Empire." *Journal of Roman Studies* 84 (1994): 1–22.

Harvey, Susan Ashbrook. "Physicians and Ascetics in John of Ephesus: An Expedient Alliance." In Scarborough 1984, 87–93.

———. *Asceticism and Society in Crisis: John of Ephesus and the Lives of the Eastern Saints.* Berkeley: University of California Press, 1990.

Hastings, James, ed. *A Dictionary of Christ and the Gospels.* 2 vols. 1906–8. Reprint, Grand Rapids, Mich.: Baker, 1973. S.v. "Gospels (Apocryphal)," 1:671–85, by A. F. Findlay.

———. *Encyclopedia of Religion and Ethics.* 13 vols. New York: Charles Scribner's Sons, 1908–26.

Hayward, Paul Antony. "Demystifying the Role of Sanctity in Western Christendom." In Howard-Johnston and Hayward 1999, 115–42.

Heil, J. P. "Significant Aspects of the Healing Miracles in Matthew." *CBQ* 41 (1979): 274–87.

Helm, Jürgen. "Sickness in Early Christian Healing Narratives: Medical, Religious and Social Aspects." In Kottek et al. 2000, 241–58.

Hemer, Colin J. *The Book of Acts in the Setting of Hellenistic History.* Edited by Conrad H. Gempf. Tübingen: J. C. B. Mohr (Paul Siebeck), 1989.

Henderson, John, Peregrine Horden, and Alessandro Pastore, eds. *The Impact of Hospitals 300–2000.* Bern: Peter Lang, 2007.

Hengel, R., and M. Hengel. "Die Heilungen Jesu und medizinisches Denken." In Suhl 1980, 338–73.

Hennecke, Edgar. *New Testament Apocrypha.* Edited by Wilhelm Schneemelcher with English translation by R. McL. Wilson. 2 vols. Philadelphia: Westminster, 1963–65.

Herzlich, Claudine, and Janine Pierret. *Illness and Self in Society.* Translated by Elborg Forster. Baltimore: Johns Hopkins University Press, 1987.

Hobart, William Kirk. *The Medical Language of St. Luke.* 1882. Reprint, Grand Rapids, Mich.: Baker, 1954.

Hogan, Larry P. *Healing in the Second Temple Period.* Novum Testamentum et Orbis Antiquus, no. 21. Freiburg, Switzerland: Universitätsverlag; Göttingen: Vandenhoeck and Ruprecht, 1992.

Holladay, A. J. "New Developments in the Problem of the Athenian Plague." *Classical Quarterly* 38 (1988): 247–50.

Holman, Susan R. "Healing the Social Leper in Gregory of Nyssa's and Gregory of Nazianzus's 'περὶ φιλοπτωχίας.'" *Harvard Theological Review* 92 (1999): 283–309.

———. "The Hungry Body: Famine, Poverty, and Identity in Basil's *Hom.* 8." *Journal of Early Christian Studies* 7, no. 3 (1999): 337–63.

———. "The Entitled Poor: Human Rights Language in the Cappadocians." *Pro Ecclesia* 9 (2000): 476–89.

——. *The Hungry Are Dying: Beggars and Bishops in Roman Cappadocia.* Oxford Studies in Historical Theology. Oxford: Oxford University Press, 2001.

Hope, V. M., and E. Marshall, eds. *Death and Disease in the Ancient City.* London: Routledge, 2000.

Hopkins, Keith. "Christian Number and Its Implications." *Journal of Early Christian Studies* 6 (1998): 185–226.

Horden, Peregrine. "Saints and Doctors in the Early Byzantine Empire: The Case of Theodore of Sykeon." In *The Church and Healing,* edited by W. J. Sheils, 1–13. Oxford: Basil Blackwell, 1982.

——. "The Byzantine Welfare State: Image and Reality." *Bulletin of the Society for the Social History of Medicine* 37 (1985): 7–10.

——. "The Death of Ascetics: Sickness and Monasticism in the Early Byzantine Middle East." In *Monastics, Hermits, and the Ascetic Tradition,* edited by W. J. Shiels, 41–52. Studies in Church History 22. Oxford: Blackwell, 1985.

——. "The Confraternities of Byzantium." In *Voluntary Religion,* edited by W. S. Sheils and Diana Wood, 25–45. Oxford: Basil Blackwell, 1986.

——. "Responses to Possession and Insanity in the Early Byzantine World." *Social History of Medicine* 6 (1993): 177–94.

——. "The Christian Hospital in Late Antiquity: Break or Bridge?" In Steger and Jankrift 2004, 77–99.

——. "The Earliest Hospitals in Byzantium, Western Europe, and Islam." *Journal of Interdisciplinary History* 35 (2005): 361–89.

——. "How Medicalised Were Byzantine Hospitals?" *Medicina e Storia* 5 (2006): 45–74.

——. "A Non-natural Environment: Medicine without Doctors and the Medieval European Hospital." In Bowers 2007, 133–45.

——. *Hospitals and Healing from Antiquity to the Later Middle Ages.* Aldershot, England: Ashgate, 2008.

Hornblower, Simon. *A Commentary on Thucydides.* Vol. I, bks. 1–3. Oxford: Clarendon Press, 1991.

Horstmanshoff, H. F. J. "The Ancient Physician: Craftsman or Scientist?" *JHM* 45 (1990): 176–97.

——. " 'Did the God Learn Medicine?' Asclepius and Temple Medicine in Aelius Aristides' *Sacred Tales.*" In Horstmanshoff and Stol 2004, 325–41.

Horstmanshoff, H. F. J., and M. Stol, eds. *Magic and Rationality in Ancient Near Eastern and Graeco-Roman Medicine.* Leiden: Brill, 2004.

Howard-Johnston, J., and P. A. Hayward, eds. *The Cult of Saints in Late Antiquity and the Middle Ages: Essays on the Contribution of Peter Brown.* Oxford: Oxford University Press, 1999.

Hull, John M. *Hellenistic Magic and the Synoptic Tradition.* London: SCM Press, 1974.

Jackson, Ralph. *Doctors and Diseases in the Roman Empire.* Norman: University of Oklahoma Press, 1988.

Jackson, Stanley W. *Melancholia and Depression from Hippocratic Times to Modern Times.* New Haven: Yale University Press, 1986.

Jacob, E. *Theology of the Old Testament.* Translated by Arthur W. Heathcote and Philip J. Allcock. New York: Harper and Row, 1958.

Jaeger, Werner. *Paideia: The Ideals of Greek Culture.* 3 vols. Translated by Gilbert Highet. New York: Oxford University Press, 1945.

——. *Early Christianity and Greek Paideia.* Cambridge: Harvard University Press, 1965.

Janowitz, Naomi. *Magic in the Roman World: Pagans, Jews, and Christians.* London: Routledge, 2001.

Jayne, W. A. *The Healing Gods of Ancient Civilizations.* New Haven: Yale University Press, 1925.

Jenkins, Claude. "Saint Augustine and Magic." In Underwood 1953, 1:131–40.

Jeremias, Joachim. *Jerusalem in the Time of Jesus: An Investigation into Economic and Social Conditions during the New Testament Period.* Translated by F. H. and C. H. Cave. 3d ed. Philadelphia: Fortress Press, 1969.

Jones, A. H. M. "The Social Background of the Struggle between Paganism and Christianity." In Momigliano 1963, 17–37.

——. *The Later Roman Empire, 284–602: A Social, Economic, and Administrative Survey.* 2 vols. Norman: University of Oklahoma Press, 1964.

Jones, C. P. "Aelius Aristides and the Asclepieion." In *Pergamon, Citadel of the Gods: Archaeological Record, Literary Description, and Religious Development,* edited by Helmut Koester, 63–76. Harrisburg, Pa.: Trinity Press International, 1998.

Jones, W. H. S., ed. and trans. *Hippocrates.* 4 vols. Loeb Classical Library. Cambridge: Harvard University Press, 1923–31.

——. *The Doctor's Oath: An Essay in the History of Medicine.* Cambridge: Cambridge University Press, 1924.

——. "Ancient Roman Folk Medicine." *JHM* 12 (1957): 459–72.

Jouanna, Jacques. *Hippocrate, Les Vents, De l'Art.* Paris: Les Belles Lettres, 1988.

——. "La Lecture de l'éthique hippocratique chez Galien." In Flashar and Jouanna 1997, 211–44.

——. *Hippocrates.* Translated by M. B. DeBevoise. Baltimore: Johns Hopkins University Press, 1999.

Judge, E. A. "The Quest for Mercy in Late Antiquity." In O'Brien and Peterson 1986, 107–21.

Kahlos, Maijestina. *Vettius Agorius Praetextatus: A Senatorial Life in Between.* Acta Instituti Romani Finlandiae 26. Rome: Institutum Romanum Finlandiae, 2002.

Kallet, L. "The Diseased Body Politic, Athenian Public Finance and the Massacre at Mykalessos (Thucydides 7.27–9)." *American Journal of Philology* 120 (1999): 223–44.

Kapparis, Konstantinos. *Abortion in the Ancient World.* London: Duckworth, 2002.

Kaufman, D. B. "Poisons and Poisoning among the Romans." *Classical Philology* 27 (1932): 156–67.

Kee, Howard Clark. *Christian Origins in Sociological Perspective: Methods and Resources.* Philadelphia: Westminster Press, 1980.

——. *Miracle in the Early Christian World: A Study in Sociohistorical Method.* New Haven: Yale University Press, 1983.

——. *Medicine, Miracle, and Magic in New Testament Times.* Cambridge: Cambridge University Press, 1986.

Keenan, Mary Emily. "Augustine and the Medical Profession." *TAPA* 67 (1936): 168–90.

——. "St. Gregory of Nazianzus and Early Byzantine Medicine." *BHM* 9 (1941): 8–30.

——. "St. Gregory of Nyssa and the Medical Profession." *BHM* 15 (1944): 150–61.

Keil, Carl Friedrich, and Franz Delitzsch. *Biblical Commentary on the Old Testament.* 25 volumes. Vol. 3, *The Pentateuch.* Translated by James Martin. 1891. Reprint, Grand Rapids, Mich.: Eerdmans, 1949.

Kelly, Henry Ansgar. *The Devil, Demonology and Witchcraft: The Development of Christian Beliefs in Evil Spirits.* Rev. ed. Garden City, N.Y.: Doubleday, 1974.

Kelly, J. N. D. *Golden Mouth: The Story of John Chrysostom, Ascetic, Preacher, Bishop.* Ithaca, N.Y.: Cornell University Press, 1995.

Kelsey, Morton T. *Healing and Christianity in Ancient Thought and Modern Times.* New York: Harper and Row, 1973.

King, Helen. "Using the Past: Nursing and the Medical Profession in Ancient Greece." In *Anthropology and Nursing,* edited by Pat Holden and Jenny Littlewood, 7–24. London: Routledge, 1991.

Kiple, Kenneth F., ed. *The Cambridge World History of Human Disease.* New York: Cambridge University Press, 1993.

Kirchner, Gernot. "Heilungswunder im Frühmittelalter: Überlegungen zum Kontext des *vir Dei*-Konzeptes Gregor von Tours." In Steger and Jankrift, 41–76.

Kislinger, Ewald. "Kaiser Julian und die (christlichen) Xenodocheia." In *Byzantios: Festschrift für Herbert Hunger zum 70. Geburtstag,* edited by W. Hörandner, J. Koder, O. Kresten, and E. Trapp, 171–84. Vienna: Ernst Becvar, 1984.

——. "Xenon und Nosokomeion." *Historia Hospitalium* 17 (1986–88): 7–16.

Kittel, Gerhard, and Gerhard Friedrich, eds. *Theological Dictionary of the New Testament.* Translated by Geoffrey W. Bromiley. 10 vols. Grand Rapids, Mich.: Eerdmans, 1964–76. S.v. "ἀγαπάω κτλ," 1:21–55, by Stauffer. S.v. "ἀλείφω," 1:229–32, by Schlier. S.v. "ἀπόστολος κτλ," 1:407–47, by Rengstorf. S.v. "δαίμων κτλ," 2:1–20, by Foerster. S.v. "εἰκών," 2:381–97, by Kittel. S.v. "ἐπιτιμάω," 2:623–27, by Stauffer. S.v. "ἰάομαι κτλ," 3:194–215, by Oepke. S.v. "φιλανθρωπία κτλ," 9:107–12, by Luck. S.v. "σατανᾶς," 7:151–63, by Foerster. S.v. "σῴζω κτλ," 7:980–1012, by Foerster.

Klawater, Frederick C. "The New Prophecy in Early Christianity: The Origin, Nature and Development of Montanism, A.D. 165–220." Ph.D. diss., University of Chicago, 1975.

Kleinman, Arthur. *Patients and Healers in the Context of Culture: An Exploration of the Borderland between Anthropology, Medicine, and Psychiatry.* Berkeley: University of California Press, 1980.

Klutz, Todd E. "What Is Specific to Western Medicine?" In Bynum and Porter 1993, 1:15–23.

——. "The Rhetoric of Science in *The Rise of Christianity*: A Response to Rodney Stark's Sociological Account of Christianization." *Journal of Early Christian Studies* 6 (1998): 162–84.

Knight, G. *A Christian Theology of the Old Testament.* Richmond, Va.: John Knox Press, 1959.

Kohn, George C. "Plague of Cyprian." In *Encyclopedia of Plague and Pestilence,* edited by George C. Kohn, 250–51. New York: Facts on File, 1995.

Kollmann, Bernd. *Jesus und die Christen als Wundertäter*: *Studien zu Magie, Medizin und Schamanismus in Antike und Christentum.* Göttingen: Vandenhoeck and Ruprecht, 1996.

Kosak, J. C. "*Polis nosousa*: Greek Ideas about the City and Disease in the Fifth Century BC." In Hope and Marshall 2000, 35–54.

Kottek, Samuel S. "Concepts of Disease in the Talmud." *Koruth* 9 (1985): 7–33.

——. *Medicine and Hygiene in the Works of Flavius Josephus.* Leiden: Brill, 1994.

——. "Hygiene and Healing among the Jews in the Post-Biblical Period: A Partial Reconstruction." In *ANRW* II. 37, 3 (1996): 2843–65.

Kottek, Samuel, Manfred Horstmanshoff, Gerhard Baader, and Gary Ferngren, eds. *From Athens to Jerusalem: Medicine in Hellenized Jewish Lore and in Early Christian Literature.* Rotterdam: Erasmus, 2000.

Kroll, Jerome. "A Reappraisal of Psychiatry in the Middle Ages." *Archives of General Psychiatry* 29 (1973): 276–83.

Kroll, Jerome, and Bernard Bachrach. "Sin and Mental Illness in the Middle Ages." *Psychological Medicine* 14 (1984): 507–14.

——. "Sin and the Etiology of Disease in pre-Crusade Europe." *JHM* 41 (1986): 395–414.

Krug, Antje. *Heilkunst und Heilkult: Medizin in der Antike.* Munich: C. H. Beck, 1985.

Kudlien, Fridolf. "Krankheitsmetaphorik des Prudentius." *Hermes* 90 (1962): 104–15.

——. "Der Arzt des Körpers und der Arzt der Seele." *Clio Medica* 3 (1968): 1–20.

——. "The Third Century A.D.—A Blank Spot in the History of Medicine?" In *Medicine, Science, and Culture: Historical Essays in Honor of Owsei Temkin,* edited by Lloyd G. Stevenson and Robert P. Multhauf, 25–34. Baltimore: Johns Hopkins Press, 1968.

——. "Medical Ethics and Popular Ethics in Greece and Rome." *Clio Medica* 5 (1970): 91–121.

——. "Cynicism and Medicine." *BHM* 48 (1974): 305–19.

——. "The Old Greek Concept of 'Relative' Health." *Journal of the History of the Behavioral Sciences* 9 (1974): 53–59.

——. "Medicine as a 'Liberal Art' and the Question of the Physician's Income." *JHM* 31 (1976): 448–59.

——. "Galen's Religious Belief." In *Galen: Problems and Prospects,* edited by Vivian Nutton, 117–30. London: Wellcome Institute, 1981.

——. "Judische Ärzte im römischen Reich." *Medizinhistorisches Journal* 20 (1985): 36–57.

Kühn, Karl Gottlob, ed. *Glaudii Galeni Opera Omnia.* 22 vols. 1821–33. Reprint, Hildesheim: Georg Olms, 1965.

Kutsch, F. *Attische Heilgötter und Heilheroen.* Diss. Religionsgeschichtliche Versuche und Vorarbeiten 12.3 (1912–13), Giessen, 1913.

Labriolle, Pierre de. *History and Literature of Christianity from Tertullian to Boethius.* Translated by Herbert Wilson. 1924. Reprint, New York: Barnes and Noble, 1968.

Ladd, George Eldon. *The New Testament and Criticism.* Grand Rapids, Mich.: Eerdmans, 1967.

——. *The Presence of the Future: The Eschatology of Biblical Realism.* 1974. Reprint, Grand Rapids, Mich.: Eerdmans, 1996.

Ladouceur, David J. "The Death of Herod the Great." *Classical Philology* 76 (1981): 25–34.

Lampe, G. W. H. "Miracles and Early Christian Apologetic." In Moule 1965, 203–18.

——. "Miracles in the Acts of the Apostles." In Moule 1965, 163–78.

——. "Diakonia in the Early Church." In McCord and Parker 1966, 49–64.

Lane Fox, Robin. *Pagans and Christians.* New York: Knopf, 1989.

Lecky, W. E. *History of European Morals: From Augustus to Charlemagne.* 2 vols. 1869. Reprint, London: Longmans Green, 1902.

Leeper, Elizabeth Ann. "Exorcism in Early Christianity." Ph.D. diss., Duke University, 1991.

Leven, Karl-Heinz. *Medizinisches bei Eusebios von Kaisareia.* Diss. Med., Düsseldorfer Arbeiten zur Geschichte der Medizin 62. Düsseldorf, 1987.

——. "Athumia and Philanthropia: Social Reactions to Plagues in Late Antiquity and Early Byzantine Society." In van der Eijk et al. 1995, 2:393–407.

——. " 'At Times These Ancient Facts Seem to Lie before Me like a Patient on a Hospital Bed'—Retrospective Diagnosis and Ancient Medical History." In Horstmanshoff and Stol 2004, 369–86.

Libanius. *Autobiography (Oration I): The Greek Text.* Edited with introduction, translation, and notes by Albert Francis Norman. London: Oxford University Press, 1965.

Lichtenthaeler, Charles. *Der Eid des Hippokrates: Ursprung und Bedeutung.* XII. Hippokratische Studie. Cologne: Deutscher Ärzte-Verlag, 1984.

Liddell, Henry George, and Robert Scott. *A Greek-English Lexicon.* New [9th] ed. Edited by Henry Stuart Jones. 1940. Oxford: Clarendon Press, 1961.

LiDonnici, Lynn R. *The Epidaurian Miracle Inscriptions: Text, Translation, and Commentary.* Atlanta: Scholars Press, 1995.

Lieber, Elinor. "Old Testament 'Leprosy,' Contagion and Sin." In Conrad and Wujastyk 2000, 99–136.

Lightfoot, John B., ed. and trans. *The Apostolic Fathers, pt. 2, S. Ignatius, S. Polycarp.* 2 vols. 1889–90. Reprint, Grand Rapids, Mich.: Baker, 1981.

Lindberg, David C. "Science and the Early Church." In *God and Nature: Historical Essays on the Encounter between Christianity and Science,* edited by David C. Lindberg and Ronald L. Numbers, 19–48. Berkeley: University of California Press, 1986.

——. "Early Christian Attitudes toward Nature." In Ferngren 2000, 243–7.

——. "Medieval Science and Religion." In Ferngren 2000, 259–67.

Lindemann, Andreas. " 'Do Not Let a Woman Destroy the Unborn Baby in Her Belly': Abortion in Ancient Judaism and Christianity." *Studia Theologica* 49 (1995): 253–71.

Little, Lester K., ed. *Plague and the End of Antiquity: The Pandemic of 541–750.* Cambridge: Cambridge University Press, 2007.

Littman, R. J., and M. L. Littman. "Galen and the Antonine Plague." *American Journal of Philology* 94 (1973): 243–55.

Littré, E. *Oeuvres Complètes d'Hippocrate.* 10 vols. Paris: J. B. Bailliere, 1839–61.

Lloyd, G. E. R. *Magic, Reason and Experience.* Cambridge: Cambridge University Press, 1979.

——. *Science, Folklore, and Ideology: Studies in the Life Sciences in Ancient Greece.* Cambridge: Cambridge University Press, 1983.

——. *Science and Morality in Greco-Roman Antiquity.* An Inaugural Lecture. Cambridge: Cambridge University Press, 1985.

——. *Demystifying Mentalities.* Cambridge: Cambridge University Press, 1990.

——. "The Invention of Nature." In *Methods and Problems in Greek Science,* 417–34. Cambridge: Cambridge University Press, 1991.

——. *In the Grip of Disease: Studies in the Greek Imagination.* Oxford: Oxford University Press, 2003.

Loewenberg, Frank M. *From Charity to Social Justice.* New Brunswick, N.J.: Transaction Publishers, 2001.

Longrigg, James. *Greek Rational Medicine: Philosophy and Medicine from Alcmaeon to the Alexandrians.* London: Routledge, 1993.

Luck, Georg. *Arcana Mundi: Magic and the Occult in the Greek and Roman Worlds.* Baltimore: Johns Hopkins University Press, 1985.

Lustig, B. Andrew. "Compassion." In Reich 1995, 1:440–5.

Macaulay, Thomas Babbington. "Life of Lord Bacon." In *Biographical Essays,* 5–153. New York: John B. Alden, 1886.

MacCasland, S. V. "Religious Healing in First-Century Palestine." In *Environmental Factors in Christian History,* edited by John Thomas MacNeill, Matthew Spinka, and Harold R. Willoughby, 27–34. Chicago: University of Chicago Press, 1939.

MacKay, B. S. "Plutarch and the Miraculous." In Moule 1965, 95–111.

MacKinney, L. C. "Medical Ethics and Etiquette in the Early Middle Ages: The Persistence of Hippocratic Ideals." *BHM* 26 (1952): 1–31.

MacMullen, Ramsay. *The Roman Government's Response to Crisis, A.D. 235–337.* New Haven: Yale University Press, 1976.

——. *Paganism in the Roman Empire.* New Haven: Yale University Press, 1981.

——. *Christianizing the Roman Empire (A.D. 100–400).* New Haven: Yale University Press, 1985.

——. "Constantine and the Miraculous." In *Changes in the Roman Empire: Essays in the Ordinary,* 107–16, 312–16. Princeton, N.J.: Princeton University Press, 1990.

——. "Two Types of Conversion to Early Christianity." In *Changes in the Roman Empire: Essays in the Ordinary,* 130–41, 322–27. Princeton, N.J.: Princeton University Press, 1990.

——. *Enemies of the Roman Order: Treason, Unrest, and Alienation in the Empire.* 1966. Reprint, London: Routledge, 1992.

——. *Christianity and Paganism in the Fourth to Eighth Centuries.* New Haven: Yale University Press, 1997.

Magoulias, H. J. "The Lives of the Saints as Sources of Data for the History of Byzantine Medicine in the Sixth and Seventh Centuries." *Byzantinische Zeitschrift* 57 (1964): 127–50.

Malina, Bruce J. *Christian Origins and Cultural Anthropology.* Atlanta: John Knox, 1986.

——. *The New Testament World: Insights from Cultural Anthropology.* 3d ed. Louisville, Ky.: Westminster John Knox, 2001.

Manchester, Keith. "Leprosy: The Origin and Development of the Disease in Antiquity." In *Maladie et maladies: Histoire et conceptualisation,* edited by Danielle Gourevitch, 31–49. Geneva: Libraire Droz, S.A., 1992.

Markus, R. A. *The End of Ancient Christianity.* Cambridge: Cambridge University Press, 1990.

Martin, Dale B. *The Corinthian Body.* New Haven: Yale University Press, 1995.

Martin, Ralph P. *2 Corinthians, Word Biblical Commentary.* Vol. 40. Waco, Tex.: Word, 1986.

Massar, Natacha. *Soigner et Servir: Histoire sociale et culturelle de la médecine grecque à l'époque hellénistique.* Paris: de Boccard, 2005.

Matousek, M. "Der Frage des Verhältnisses der Urchristentums zur Medicine." *Zeitschrift für Geschichte der Naturwissenschaften. Technik und Medizin* 1 (1960), H. 3: 74–9.

Mayor, Joseph B. *The Epistle of St. James: The Greek Text with Introduction.* 3d ed. 1913. Reprint, Grand Rapids, Mich.: Zondervan, 1954.

McCord, James I., and T. H. L. Parker. *Service to Christ: Essays Presented to Karl Barth on his 80th Birthday.* Grand Rapids, Mich.: Eerdmans, 1966.

McEwen, J. S. "The Ministry of Healing." *SJT* 7 (1954): 133–52.

McMillan, R. M., H. T. Englehardt Jr., and S. F. Spicker, eds. *Euthanasia and the Newborn: Conflicts Regarding Saving Lives.* Dordrecht: Reidel, 1987.

McNeill, John T. *A History of the Cure of Souls.* New York: Harper and Row, 1951.

Meeks, Wayne A. *The First Urban Christians: The Social World of the Apostle Paul.* New Haven: Yale University Press, 1983.

Meer, Frederic van der. *Augustine the Bishop: The Life and Work of a Father of the Church.* Translated by B. Battershaw and G. R. Lamb. London: Sheed and Ward, 1961.

Meier, Mischa. "Von Prokop zu Gregor von Tours: Kultur- und mentalitätengeschichtliche relevante Folgen der 'Pest' im 6. Jahrhundert." In Steger and Jankrift 2004, 19–40.

Melinsky, M. A. H. *Healing Miracles.* London: Mowbray, 1968.

Merideth, Anne Elizabeth. "Illness and Healing in the Early Christian East." Ph.D. diss., Princeton University, 1999.

Miles, Margaret. *Fullness of Life: Historical Foundations for a New Asceticism.* Philadelphia: Westminster, 1981.

Miller, Harold W. "The Concept of the Divine in De Morbo Sacro." *TAPA* 84 (1953): 1–15.

Miller, J. M. "In the 'Image' and 'Likeness' of God." *Journal of Biblical Literature* 91 (1972): 289–304.

Miller, Timothy S. *The Birth of the Hospital in the Byzantine Empire.* 2d ed. Baltimore: Johns Hopkins University Press, 1997.

Minnen, Peter van. "Medical Care in Late Antiquity." In van der Eijk et al. 1995, 1:153–69.

Mitchell-Boyask, Robin. *Plague and the Athenian Imagination: Drama, History and the Cult of Asclepius.* Cambridge: Cambridge University Press, 2008.

Molland, E. "Ut sapiens medicus: Medical Vocabulary in St. Benedikt's Regula Monachorum." *Studia Monastica* 6 (1964): 273–96.

Momigliano, Arnaldo, ed. *The Conflict between Paganism and Christianity in the Fourth Century.* Oxford: Clarendon Press, 1963.

——. "Popular Religious Beliefs and the Late Roman Historians." In *Popular Belief and Practice,* edited by G. J. Cuming and Derek Baker, 1–18. Studies in Church History 8. Cambridge: Cambridge University Press, 1972.

Morgan, Thomas E. "Plague or Poetry? Thucydides on the Epidemic at Athens." *TAPA* 124 (1994): 197–209.

Morrison, E. F. *St. Basil and His Rule: A Study in Early Monasticism.* London: Oxford University Press, 1912.

Moule, C. F. D., ed. *Miracles: Cambridge Studies in Their Philosophy and History.* London: Mowbray, 1965.

——. "The Vocabulary of Miracle." In Moule 1965, 235–8.

Mudry, Philippe. "Éthique et médecine à Rome: La Préface de Scribonius Largus ou l'affirmation d'une singularité." In Flashar and Jouanna 1997, 297–322.

Mudry, Philippe, and Pigeaud, Jackie, eds. *Les Écoles médicales à Rome.* Geneva: Librairie Droz, 1991.

Müller, Gerhard. "Arzt, Kranker und Krankheit bei Ambrosius von Mailand (334–97)." *Sudhoffs Archiv* 51 (1967): 193–216.

Natali, A. "Eglise et évergétisme à Antioche à la fin du IVe siècle d'après Jean Chrysostome." *Studia Patristica* 17 (1982): 1176–84.

Naveh, Joseph. "A Medical Document or a Writing Exercise? The So-called 4Q Therapeia." *Israel Exploration Journal* 36 (1986): 52–5.

Neusner, J., ed. *Christianity, Judaism, and Other Greco-Roman Cults: Studies for Morton Smith at Sixty.* Part 1: *New Testament.* Leiden: Brill, 1975.

Newmyer, Stephen T. "Talmudic Medicine and Greco-Roman Science: Crosscurrents and Resistance." In *ANRW* II. 37, 3 (1996): 2895–2911.

Nicholson, O. P. "The Date of Arnobius' *Adversus gentes.*" *Studia Patristica* 15 (1984): 101–7.

Nilsson, Martin Persson. *Greek Piety.* Translated by Herbert Jennings Rose. Oxford: Clarendon Press, 1948.

Nock, Arthur D. "Paul and the Magus." In *The Beginnings of Christianity,* edited by F. Jackson and V. Laue, 164–88. Cambridge: Harvard University Press, 1933.

——. *Conversion: The Old and the New in Religion from Alexander the Great to Augustine of Hippo.* Oxford: Clarendon Press, 1933.

Noonan, John T., Jr. "An Almost Absolute Value in History." In *The Morality of Abortion: Legal and Historical Perspectives,* edited by John T. Noonan Jr., 51–9. Cambridge: Harvard University Press, 1970.

——. *Contraception: A History of Its Treatment by the Catholic Theologians and Canonists.* Cambridge: Harvard University Press, 1986.

Noorda, Sijbolt J. "Illness and Sin, Forgiving and Healing: The Connection of Medical Treatment and Religious Beliefs in Sira 38, 1–15." In *Studies in Hellenistic Religions,* edited by Maarten Jozef Vermaseren, 215–24. Leiden: Brill, 1979.

Nowak, Edward. *Le Chrétien devant la souffrance: Étude sur la pensée de Jean Chrysostome.* Theol. Crit. 19. Paris: Beauchesne, 1973.

Nussbaum, Martha C. *The Therapy of Desire: Theory and Practice in Hellenistic Ethics.* Princeton, N.J.: Princeton University Press, 1994.

Nutton, Vivian. "Two Notes on Immunities: Digest 27, 1, 6, 10, and 11." *Journal of Roman Studies* 61 (1971): 52–63. Reprinted in Nutton 1988, IV 52–63.

——. "Museums and Medical Schools in Classical Antiquity." *History of Education* 4 (1975): 3–15.

——. "Continuity or Rediscovery? The City Physician in Classical Antiquity and Mediaeval Italy." In *The Town and State Physician in Europe from the Middle Ages to the Enlightenment,* edited by Andrew W. Russell, 9–46. Wolfenbütter: Herzog August Bibliothek, 1981. Reprinted in Nutton 1988, VI 9–46.

——. "The Seeds of Disease: An Explanation of Contagion and Infection from the Greeks to the Renaissance." *Medical History* 27 (1983): 1–34. Reprinted in Nutton 1988, XI 1–34.

——. "From Galen to Alexander: Aspects of Medicine and Medical Practice in Late Antiquity." In Scarborough 1984, 1–14.

——. "Murders and Miracles: Lay Attitudes towards Medicine in Classical Antiquity." In *Patients and Practitioners: Lay Perceptions of Medicine in Pre-Industrial Society,* edited by Roy Porter, 45–51. Cambridge: Cambridge University Press, 1985. Reprinted in Nutton 1988, VIII 23–53.

——. Essay Review of *The Birth of the Hospital in the Byzantine Empire* by Timothy S. Miller. *Medical History* 30 (1986): 218–21.

——. *From Democedes to Harvey: Studies in the History of Medicine.* London: Variorum, 1988.

——. "From Medical Certainty to Medical Amulets: Three Aspects of Ancient Therapeutics." In *Essays in the History of Therapeutics*, edited by W. F. Bynum and Vivian Nutton, 13–22. Amsterdam: Rodopi, 1991.

——. "Healers in the Medical Market Place: Towards a Social History of Graeco-Roman Medicine." In *Medicine and Society*, edited by Andrew Wear, 1–58. Cambridge: Cambridge University Press, 1992.

——. "Beyond the Hippocratic Oath." In *Doctors and Ethics: The Earlier Historical Setting of Professional Ethics*, edited by Andrew Wear, Johanna Geyer-Kordesch, and Roger French, 19–37. Amsterdam: Rodopi, 1993.

——. "Roman Medicine: Tradition, Confrontation, Assimilation." In *ANRW* II. 37, 1 (1993): 49–78.

——. "Humoralism." In Bynum and Porter 1993, 1:281–91.

——. "The Medical Meeting Place." In van der Eijk et al. 1995, 1:3–25.

——. "Hippocratic Morality and Modern Medicine." In Flashar and Jouanna 1997, 31–56.

——. "Did the Greeks Have a Word for It? Contagion and Contagion Theory in Classical Antiquity." In Conrad and Wujastyk 2000, 137–62.

——. "God, Galen and the Depaganization of Ancient Medicine." In *Religion and Medicine in the Middle Ages*, edited by Peter Biller and Joseph Ziegler, 17–32. York: York Medieval Press, 2001.

——. Review of *From Athens to Jerusalem: Medicine in Hellenized Jewish Lore and in Early Christian Literature* by S. Kottek et al. *BHM* 75 (2001): 787–8.

——. *Ancient Medicine*. London: Routledge, 2004.

Oates, W. J., ed. *The Stoic and Epicurean Philosophers*. 1940. New York: Modern Library, 1957.

O'Brien, P. T., and D. G. Peterson, eds. *God Who Is Rich in Mercy: Essays Presented to D. B. Knox*. Homebush West, NSW, Australia: Anzear, 1986.

Oliver, James H. "Two Athenian Poets." *Hesperia*, Supplement 8 (1949): 243–58.

Oliver, James H., and Paul L. Maas. "An Ancient Poem on the Duties of a Physician." *BHM* 7 (1939): 315–23.

Palmer, Bernard, ed. *Medicine and the Bible*. Carlisle, England: Paternoster Press, 1986.

Papagrigorakis, Manolis, Christos Yapijakis, Phillippos Synodinos, and Effie Baziotapoulou-Valavani. "DNA Examination of Ancient Dental Pulp Incriminates Typhoid Fever as Probable Cause of the Plague of Athens." *International Journal of Infectious Diseases* 10 (2006): 206–14.

Parker, Robert. *Miasma: Pollution and Purification in Early Greek Religion*. Oxford: Clarendon Press, 1983.

Parry, Adam. "The Language of Thucydides' Description of the Plague." *Bulletin of the Institute of Classical Studies* 16 (1969): 106–18.

Patlagean, Evelyne. *Pauvreté économique et pauvreté sociale à Byzance 4e–7e siècles*. Paris: Mouton, 1977.

——. "The Poor." In *The Byzantines*, edited by Guglielmo Cavallo, translated by Thomas Dunlap, Teresa Lavender Fagan, and Charles Lambert, 15–42. Chicago: University of Chicago Press, 1997.

Pattengale, Jerry A. "Benevolent Physicians in Late Antiquity: The Cult of the Anargyroi." Ph.D. diss., Miami University, 1993.

Paxton, Frederick S. *Christianizing Death: The Creation of a Ritual Process in Early Medieval Europe.* Ithaca, N.Y.: Cornell University Press, 1990.

Pease, Arthur Stanley. "Medical Allusions in the Works of St. Jerome." *Harvard Studies in Classical Philology* 25 (1914): 73–86.

Perkins, Judith. *The Suffering Self: Pain and Narrative Representation in the Early Christian Era.* London: Routledge, 1995.

Petersen, J. "Dead or Alive? The Holy Man as Healer in East and West in the Late Sixth Century." *Journal of Medieval History* 9 (1983): 91–8.

Pétré, Hélène. *Caritas: Étude sur le vocabulaire latin de la charité chrétienne.* Louvain: Spicilegium Sacrum Lovaniense, 1948.

Pétridès, S. "Spoudaei et philopones." *Echos d'Orient* 7 (1904): 341–8.

Pharr, Clyde. "The Interdiction of Magic in Roman Law." *TAPA* 63 (1932): 269–95.

Philipsborn, A. "La Compagnie d'Ambulanciers 'Parabalini' d'Alexandrie." *Byzantion* 20 (1950): 185–90.

Phillips, Joanne H. "The *Liber Medicinalis Quinti Sereni* and Popular Medicine: A Reconsideration of Sources Required." In Mudry and Pigeaud 1991, 179–86.

Pigeaud, Jackie. *La Maladie de l'ame: Étude sur la relation d l'âme et du corps dans la tradition médico-philosophique antique.* Paris: Société d'Édition "Les Belles Lettres," 1981.

——. "Les Fondements philosophiques de l'éthique médicale: Le cas de Rome." In Flashar and Jouanna 1997, 255–96.

Pilch, John J. *Healing in the New Testament: Insights from Medical and Mediterranean Anthropology.* Minneapolis: Fortress, 2000.

Pleket, H. W. "The Social Status of Physicians in the Graeco-Roman World." In van der Eijk et al. 1995, 1:27–34.

Poole, J. C. F., and A. J. Holladay, "Thucydides and the Plague of Athens." *Classical Quarterly* 29 (1979): 282–300.

Porteous, N. W. "Image of God." In *The Interpreter's Dictionary of the Bible,* edited by G. A. Buttrick, 2:682–5. Nashville: Abingdon, 1962.

——. "The Care of the Poor in the Old Testament." In McCord and Parker 1966, 27–37.

Praet, Danny. "Explaining the Christianization of the Roman Empire: Older Theories and Recent Developments." *Sacris Erudiri: Jaarboek voor Godsdienstwetenschappen* 33 (1992–93): 7–119.

Preuss, J. *Julius Preuss' Biblical and Talmudic Medicine.* Translated by Fred Rosner. New York: Hebrew Publishing, 1978.

Price, Robert M. "Illness Theodicies in the New Testament." *Journal of Religion and Health* 25 (1986): 309–15.

Prioreschi, P. "Did Hippocratic Physicians Treat Hopeless Cases?" *Gesnerus* 49 (1992) (part 3/4): 341–9.

Quasten, Johannes. *Patrology.* 4 vols. Westminster, Md.: Christian Classics, 1950–86.

Ranger, T., and P. Slack, eds. *Epidemics and Ideas: Essays on the Historical Perception of Pestilence.* Cambridge: Cambridge University Press, 1992.

Ratzan, R. M., and Gary B. Ferngren. "A Greek Progymnasma on the Physician-Poisoner." *JHM* 48 (1993): 157–70.

Rebillard, Éric. "Église et sépulture dans l'Antiquité tardive (Occident latin, 3e–6e siècles)." *Annales HSS* 54, no. 5 (1999): 1027–46.

——. "La 'Conversion' de l'Empire romain selon Peter Brown (note critique)." *Annales HSS* 54, no. 4 (1999): 813–23.

——. "Les Formes de l'assistance funéraire dans l'empire romain et leur évolution dans l'antiquité tardive." *Antiquité tardive* 7 (1999): 269–82.

Reff, Daniel T. *Plagues, Priests, and Demons: Sacred Narratives and the Rise of Christianity in the Old World and the New.* Cambridge: Cambridge University Press, 2005.

Reich, Warren T., ed. *The Encyclopedia of Bioethics.* 2d ed. 5 vols. New York: Simon and Schuster Macmillan, 1995.

Remus, Harold. "Does Terminology Distinguish Early Christian from Pagan Miracles?" *Journal of Biblical Literature* 101 (1982): 531–51.

——. "Magic or Miracle? Some Second Century Instances." *Second Century* 2 (1983): 127–56.

——. *Pagan-Christian Conflict over Miracle in the Second Century.* Cambridge: Philadelphia Patristic Foundation, 1983.

——. *Jesus as Healer.* New York: Cambridge University Press, 1997.

Rialdi, G. *La medicina nella dottrina di Tertulliano.* Scientia Veterum 126. Pisa: Casa Editrice Giardini, 1968.

Richardson, Alan. *The Miracle Stories of the Gospels.* London: SCM Press, 1941.

Riddle, John M. "Folk Tradition and Folk Medicine: Recognition of Drugs in Classical Antiquity." In Scarborough 1987, 33–61.

——. *Eve's Herbs: A History of Contraception and Abortion in the West.* Cambridge: Harvard University Press, 1997.

——. "Research Procedures in Evaluating Medieval Medicine." In Bowers 2007, 3–17.

Riethmüller, Jürgen W. *Asklepios: Heiligtümer und Kulte.* Heidelberg: Verlag Archäologie und Geschichte, 2005.

Risse, Guenter B. *Mending Bodies, Saving Souls: A History of Hospitals.* New York: Oxford University Press, 1999.

Rist, J. M. *Human Value: A Study in Ancient Philosophical Ethics.* Leiden: Brill, 1982.

Roberts, A., and J. Donaldson, eds. *The Ante-Nicene Fathers.* 10 vols. 1885–96. Reprint, Grand Rapids, Mich.: Eerdmans, 1956.

Robinson, H. W. *The Christian Doctrine of Man.* Edinburgh: T. and T. Clark, 1926.

Rosner, Fred. "Jewish Medicine in the Talmudic Period." In *ANRW* II. 37, 3 (1996): 2866–94.

Rousseau, Philip. *Basil of Caesarea.* The Transformation of the Classical Heritage 20. Berkeley: University of California Press, 1994.

Rousselle, Aline. "From Sanctuary to Miracle-Worker: Healing in Fourth-Century Gaul." In *Ritual, Religion, and the Sacred: Selections from the Annales, Économies, Sociétés, Civilisations,* vol. 7, edited by R. Forster and O. Ranum, translated by E. Forster and P. M. Ranum, 95–127. Baltimore: Johns Hopkins University Press, 1982. English translation of "Du Sanctuaire au thaumaturge: La guérison en gaule au IVe siècle." *Annales, économies, sociétés, civilisations* 31 (1976): 1085–1107.

——. *Porneia: On Desire and the Body in Antiquity.* Translated by Felicia Pheasant. Oxford: Basil Blackwell, 1988.

——. *Croire et guérir: La foi en Gaule dans l'Antiquité tardive.* Paris: Fayard, 1990.

Russell, Colin A. "The Conflict of Science and Religion." In Ferngren 2000, 12–16.

Russell, Jeffrey Burton. *The Devil: Perceptions of Evil from Antiquity to Primitive Christianity.* Ithaca, N.Y.: Cornell University Press, 1977.

——. *Satan: The Early Christian Tradition.* Ithaca, N.Y.: Cornell University Press, 1981.

Rütten, Thomas. "Medizenethische Themen in den deontologischen Schriften des *Corpus Hippocraticum.*" In Flashar and Jouanna 1997, 65–111.

——. *Geschichten vom Hippokratischen Eid.* Electronic resource. Wiesbaden: Harrossowitz in Kommission, 2007.

Saller, Benson. "Supernatural as a Western Category." *Ethos* 5 (1977): 31–53.

Sallares, R. *Malaria and Rome: A History of Malaria in Ancient Italy.* Oxford: Oxford University Press, 2002.

Scarborough, John, ed. *Symposium on Byzantine Medicine.* Dumbarton Oaks Papers no. 38. Washington, D.C.: Dumbarton Oaks, 1984.

——. "Adaptation of Folk Medicines in the Formal Materia Medica of Classical Antiquity." In Scarborough 1987, 21–32.

——. *Folklore and Folk Medicines.* Madison, Wis.: American Institute of the History of Pharmacy, 1987.

Schadewaldt, Hans. "Die Apologie der Heilkunst bei den Kirchenvätern." *Veröffentlichungen der Internationalen Gesellschaft für Geschichte der Pharmazie* 26 (1965): 115–30.

Schipperges, H. "Zur Tradition des 'Christus Medicus' im frühen Christentum und in der älteren Heilkunde." *Arzt und Christ* 11 (1965): 12–20.

Schubart, W. "Parabalani." *Journal of Egyptian Archaeology* 40 (1954): 97–101.

Schubert, Charlotte. *Der hippokratische Eid: Medizin und Ethik von der Antike bis heute.* Darmstadt: Wissenschaftliche Buchgesellschaft, 2005.

Schulze, Christian. "Christliche Ärztinnen in der Antike." In Schulze and Ihm 2002, 91–115.

——. *Medizin und Christentum in Spatäntike und frühem Mittelalter: Christliche Ärzte und ihr Wirken.* Tübingen: Mohr Siebeck, 2005.

Schulze, Christian, and Sibylle Ihm, eds. *Ärztekunst und Gottvertrauen: Antike und mittelalterliche Schittpunkte von Christentum und Medizin.* Spudasmata Band 86. Hildesheim: Georg Olms, 2002.

Schweikardt, Christoph, and Christian Schulze. "Facetten antiker Krankenpflege und ihrer Rezeption." In Schulze and Ihm 2002, 117–38.

Scobie, Alex. "Slums, Sanitation, and Mortality in the Roman World." *Klio* 68 (1986): 399–433.

Sconocchia, Sergio, ed. *Scribonii Largi Compositiones.* Leipzig: Teubner, 1983.

——. "Le Problème des sectes médicales à Rome au Ier s. ap. J.-C. d'apres l'oeuvre de Scribonius Largus." In Mudry et Pigeaud 1991, 138–47.

Seccombe, David. "Was There Organized Charity in Jerusalem before the Christians?" *Journal of Theological Studies* 29 (1978): 140–3.

Segal, J. B. *Edessa "The Blessed City."* Oxford: Clarendon Press, 1970.

Seneca. *Moral and Political Essays.* Edited and translated by John M. Cooper and J. F. Procopé. Cambridge: Cambridge University Press, 1995.

Seybold, Klaus, and Ulrich B. Mueller. *Sickness and Healing.* Translated by Douglas W. Stott. Nashville: Abingdon, 1981.

Shelp, Earl E., ed. *Virtue and Medicine: Explorations in the Character of Medicine.* Dordrecht: Reidel, 1985.

Sherwin-White, A. N. *Roman Society and Roman Law in the New Testament.* Oxford: Clarendon Press, 1963.

Sigerist, Henry E. *Civilization and Disease.* Chicago: University of Chicago Press, 1943.

——. *A History of Medicine.* 2 vols. London: Oxford University Press, 1951–61.

——. "The Special Position of the Sick." In *Culture, Disease, and Healing,* edited by David Landy, 388–94. New York: Macmillan, 1977.

Simmons, Michael Bland. *Arnobius of Sicca: Religious Conflict and Competition in the Age of Diocletian.* Oxford: Clarendon Press, 1995.

Simon, Bennett. *Mind and Madness in Ancient Greece: The Classical Roots of Modern Psychiatry.* Ithaca, N.Y.: Cornell University Press, 1978.

Skemp, J. B. "Service to the Needy in the Graeco-Roman World." In McCord and Parker 1966, 17–26.

Smith, Jonathan Z. "Good News Is No News: Aretalogy and Gospel." In *Christianity, Judaism and Other Greco-Roman Cults: Studies for Morton Smith at Sixty,* edited by Jacob Neusner, 21–38. Leiden: Brill, 1975.

Smith, Morton. "Prolegomena to a Discussion of Aretalogies, Divine Men, the Gospels and Jesus." *Journal of Biblical Literature* 90 (1971): 174–99.

——. "De tuenda sanitate praecepta (Moralia, 122B–137E)." In *Plutarch's Ethical Writings and Early Christian Literature,* edited by Hans Dieter Betz, 32–50. Leiden: Brill, 1978.

——. *Jesus the Magician.* New York: Harper and Row, 1978.

Smith, Wesley D. "So-Called Possession in Pre-Christian Greece." *TAPA* 96 (1965): 403–26.

——, ed. and trans. *Hippocrates: Pseudepigraphic Writings.* Leiden: Brill, 1990.

Smith, William, and Samuel Cheetham. *A Dictionary of Christian Antiquities.* 2 vols. London: John Murray, 1875–80. S.v. "Hospitals," 1:785–9. S.v. "Exorcism," 2:650–3. S.v. "Unction," 2:2000–6. S.v. "Wonders," 2:2041–54.

Smith, William, and Henry Wace. *A Dictionary of Christian Biography.* 4 vols. Boston: Little, Brown, 1877–87. S.v. "Cyprianus (1) Thascius Caecilius," 1:739–55.

Snell, B. *The Discovery of the Mind.* Translated by T. G. Rosenmeyer. New York: Harper and Row, 1960.

Stark, Rodney. "Epidemics, Networks, and the Rise of Christianity." *Semeia* 56 (1992): 159–75.

——. *The Rise of Christianity: A Sociologist Reconsiders History.* Princeton, N.J.: Princeton University Press, 1996.

——. "E Contrario." *Journal of Early Christian Studies* 6 (1998): 259–67.

Steger, Florian. *Asklepiosmedizin: Medizinischer Alltag in der römischen Kaiserzeit.* Stuttgart: Franz Steiner Verlag, 2004.

Steger, Florian, and Kay Peter Jankrift, eds. *Gesundheit—Krankheit: Kulturtransfer medizinischen Wissens von der Späntantike bis in die Frühe Neuzeit.* Koln: Böhlau, 2004.

Steidle, P. Basilius. "Ich war krank, und ihr habt mich besucht' (Mt 25, 36)." *Erbe und Auftrag* 40 (1964): 443–58; 41 (1965): 36–46, 99–113, 189–206.

Sternberg, Thomas. *Orientalium More Secutus: Räume und Institutionen der Caritas des 5. bis 7. Jahrhunderts in Gallien. Jahrbuch für Antike und Christentum Ergänzungsband* 16. Münster: Aschendorffsche, 1991.

Strenski, Ivan. "Religon, Power, and Final Foucault." *Journal of the American Academy of Religion* 66, no. 2 (1998): 345–67.

Stroumsa, Gedaliahu G. "The Manichaean Challenge to Egyptian Christianity." In *The Roots of Egyptian Christianity*, edited by Birger Pearson and James E. Goehring, 307–19. Philadelphia: Fortress, 1986.

——. *"Caro salutis cardo:* Shaping the Person in Early Christian Thought." *History of Religions* 30 (1990): 86–99.

Suhl, A., ed. *Der Wunderbegriff im Neuen Testamentum.* Darmstadt: Wissenschaftliche Buchgesellschaft, 1980.

Swain, Simon. "Man and Medicine in Thucydides." *Arethusa* 27 (1994): 303–28.

Tarn, W. W., and G. T. Griffith. *Hellenistic Civilization.* 3d ed. London: Arnold, 1952.

Temkin, Owsei. *Soranus' Gynecology.* Translated with an introduction by Owsei Temkin. 1956. Reprint, Baltimore: Johns Hopkins University Press, 1991.

——. *The Falling Sickness: A History of Epilepsy from the Greeks to the Beginnings of Modern Neurology.* 2d ed, rev. Baltimore: Johns Hopkins Press, 1971.

——. *Galenism: Rise and Decline of a Medical Philosophy.* Ithaca, N.Y.: Cornell University Press, 1973.

——. *The Double Face of Janus and Other Essays in the History of Medicine.* Baltimore: Johns Hopkins University Press, 1977.

——. "The Scientific Approach to Disease: Specific Entity and Individual Sickness." In Temkin 1977, 441–55.

——. "Medical Ethics and Honoraria in Late Antiquity." In *Healing and History: Essays for George Rosen*, edited by Charles E. Rosenberg, 6–26. New York: Science History Publications, 1979.

——. *Hippocrates in a World of Pagans and Christians.* Baltimore: Johns Hopkins University Press, 1991.

——. *"On Second Thought" and Other Essays in the History of Medicine and Science.* Baltimore: Johns Hopkins University Press, 2002.

Temporini, Hildegard, and Wolfgang Haase, eds. *Aufstieg und Niedergang der Römischen Welt.* Berlin: Walter de Gruyter, 1972–.

Thielman, Samuel B., and Frank S. Thielman. "Constructing Religious Melancholy: Despair, Melancholia, and Spirituality in the Writings of the Church Fathers of Late Antiquity." Paper presented at the sixty-ninth meeting of the American Association for the History of Medicine, Williamsburg, Va., April 4, 1997.

Theissen, Gerd. *Sociology of Early Palestinian Christianity.* Translated by John Bowden. Philadelphia: Fortress Press, 1978.

——. *The Miracle Stories of the Early Christian Tradition.* Translated by F. McDonagh. Philadelphia: Fortress Press, 1983.

Thrämer, E. "Health and Gods of Healing (Greek)." In *ERE* 6:540–53.

——. "Health and Gods of Healing (Roman)." In *ERE* 6:553–6.

Thucydides. *History of the Peloponnesian War.* Translated by Rex Warner with an introduction and notes by M. I. Finley. London: Penguin, 1974.

Tiede, David Lenz. *The Charismatic Figure as Miracle Worker.* Missoula, Mont.: University of Montana, 1972.

Trevett, Christine. *Montanism: Gender, Authority and the New Prophecy.* Cambridge: Cambridge University Press, 1996.

Turner, H. E. W. *The Patristic Doctrine of Redemption: A Study of the Development of Doctrine during the First Five Centuries.* London: Mowbray, 1952.

Twelftree, Graham H. *Jesus the Exorcist: A Contribution to the Study of the Historical Jesus.* Tübingen: Mohr, 1993.

——. *Jesus the Miracle Worker: A Historical and Theological Study.* Downers Grove, Ill.: InterVarsity Press, 1999.

Uhlhorn, Gerhard. *Christian Charity in the Ancient Church.* Translated from the German. New York: Charles Scribner's Sons, 1883.

Underwood, Edgar Ashworth, ed. *Science, Medicine, and History: Essays on the Evolution of Scientific Thought and Medical Practice Written in Honour of Charles Singer.* 2 vols. London: Oxford University Press, 1953.

Uytfanghe, M. van. "La Controverse biblique et patristique autour du miracle, et ses répercussions sur l'hagiographie dans l'Antiquité tardive et le haute Moyen Âge latin." In *Hagiographie, cultures et sociétés (IV–XII siécles)* 205–33. Paris: Études Augustiniennes, 1981.

Vaage, Leif E., and Vincent L. Wimbrush, eds. *Asceticism and the New Testament.* London: Routledge, 1999.

Van Dam, Raymond. "Hagiography and History: The Life of Gregory Thaumaturgus." *Classical Antiquity* 1 (1982): 272–308.

——. *Saints and Their Miracles in Late Antique Gaul.* Princeton, N.J.: Princeton University Press, 1993.

van der Eijk, Philip J. "The 'Theology' of the Hippocratic Treatise, *On the Sacred Disease.*" In van der Eijk 2005, 45–73.

——. *Medicine and Philosophy in Classical Antiquity: Doctors and Philosophers on Nature, Soul, Health and Disease.* New York: Cambridge University Press, 2005.

van der Eijk, Philip J., H. F. J. Horstmanshoff, and P. H. Schrijvers, eds. *Ancient Medicine in Its Socio-Cultural Context.* Clio Medica 27. 2 vols. Amsterdam: Rodopi, 1995.

Van der Horst, P. W. *Aelius Aristides and the New Testament.* Studia ad Corpus Hellenisticum Novi Testamenti, vol. 6. Leiden: Brill, 1980.

Van der Loos, Hendrik. *The Miracles of Jesus.* Translated by T. S. Preston. Nov. Test. Supple. 9. Leiden: Brill, 1965.

van der Toorn, K. *Sin and Sanction in Israel and Mesopotamia: A Comparative Study.* Assen, Netherlands: Van Gorcum, 1985.

Vermès, Géza. *Jesus the Jew: A Historian's Reading of the Gospels.* London: Collins, 1973.

Veyne, Paul. *Bread and Circuses: Historical Sociology and Political Pluralism.* Translated by Brian Pearce. London: Penguin, 1990.

Vlastos, Gregory. "Religion and Medicine in the Cult of Asclepius: A Review Article." *Review of Religion* 13 (1949): 269–90.

Von Allmen, J.-J., ed. *A Companion to the Bible.* New York: Oxford University Press, 1958. S.v. "Sickness," 402–4, by H. Roux. S.v. "Victory," 433–5, by H. Roux.

von Nordheim, M. *"Ich bin der Herr, dein Arzt": Der Arzt in der Kultur des alten Israel. Würzburger medizinhistorische Forschungen,* Bd. 63. Edited by G. Keil. Würzburg: Königshausen & Neuman, 1998.

von Rad, Gerhard. *Old Testament Theology.* Vol. I. Translated by D. M. G. Stalker. New York: Harper and Row, 1962.

Von Staden, Heinrich. "Incurability and Hopelessness: The Hippocratic Corpus." In *La Maladie et les maladies dans la Collection hippocratique,* edited by P. Potter, G. Maloney, and J. Desautels, 75–112. Quebec: Editions du Sphinx, 1990.

——. "In a Pure and Holy Way: Personal and Professional Conduct in the Hippocratic Oath." *JHM* 51 (1996): 404–37.

——. "Character and Competence: Personal and Professional Conduct in Greek Medicine." In Flashar and Jouanna 1997, 157–95.

Vööbus, Arthur. *History of Asceticism in the Syrian Orient: A Contribution to the History of Culture in the Near East.* Vol. 1, *The Origin of Asceticism: Early Monasticism in Persia.* Louvain: Secrétariat du CorpusSCO, 1958.

Wallace-Hadrill, D. *The Greek Patristic View of Nature.* Manchester: Manchester University Press, 1968.

Walzer, Richard. *Galen on Jews and Christians.* Oxford: Oxford University Press, 1949.

——. "New Light on Galen's Moral Philosophy (from a Recently Discovered Arabic Source)." *Classical Quarterly* 43 (1949): 82.

Ward, Benedicta. *Miracles and the Medieval Mind: Theory, Record, and Event, 1000–1215.* Rev. ed. Aldershot, England: Scolar Press, 1987.

Warfield, Benjamin B. *Counterfeit Miracles.* 1918. Reprint, London: Banner of Truth, 1972.

Waszink, Jan Hendrik. *Tertullian. De anima.* Edited with introduction and commentary. Amsterdam: J. M. Meulenhoff, 1947.

Weinreich, O. *Antike Heilungswunder: Untersuchungen zum Wunderglauben der Griechen und Römer. Religionsgeschichtliche Versuche und Vorarbeiten* 8, Bd. 1. Giessen: Topelmann, 1909.

White, L. Michael. "Adolf Harnack and the Expansion of Early Christianity." *Second Century* 5 (1985/86): 97–127.

Whitehorne, J. E. G. "Was Marcus Aurelius a Hypochondriac?" *Latomus* 36 (1977): 413–41.

Wiles, M. F. "Miracles in the Early Church." In Moule 1965, 219–34.

Wilken, Robert L. *The Christians As the Romans Saw Them.* New Haven: Yale University Press, 1984.

Wilkinson, John. "A Study of Healing in the Gospel According to John." *SJT* 20 (1965): 442–61.

——. "Healing in the Epistle of James." *SJT* 24 (1971): 326–45.

Wilson, David B. "The Historiography of Science and Religion." In Ferngren 2000, 3–11.

Winslow, Donald F. "Gregory of Nazianzus and Love for the Poor." *Anglican Theological Review* 47 (1965) 348–59.

Wipszycka, Ewa. "Les confréries dans la vie religieuse de l'Egypte chrétienne." In *Proceedings of the Twelfth International Congress of Papyrology,* edited by Deborah H. Samuel, 511–25. American Studies in Papyrology, vol. 7. Toronto: A. M. Hakkert, 1970.

Witt, R. E. *Isis in the Graeco-Roman World.* Ithaca, N.Y.: Cornell University Press, 1971.

Wolff. H. W. *Anthropology of the Old Testament.* Translated by Margaret Kohl. Philadelphia: Fortress Press, 1974.

Wright, G. Ernest, and Reginald H. Fuller. *The Books of the Acts of God: Contemporary Scholarship Interprets the Bible.* Garden City, N.Y.: Doubleday, Anchor, 1960.

Wright, N. T. *The New Testament and the People of God.* Vol. 1 of *Christian Origins and the Question of God.* Minneapolis: Fortress Press, 1992.

——. *Jesus and the Victory of God.* Vol. 2 of *Christian Origins and the Question of God.* Minneapolis: Fortress Press, 1992.

Yamauchi, Edwin. "Magic or Miracle? Diseases, Demons and Exorcism." In *The Miracles of Jesus,* edited by David Wenham and Craig Blomberg, 89–183. Gospel Perspectives 6. Sheffield, England: JSOT Press, 1986.

INDEX

Under the entry for *Bible,* page numbers are in *italic* to distinguish them from chapter and verse numbers.